Vitus B. Dröscher

The Magic of the Senses

new discoveries in animal perception

D1795605

Panther

Granada Publishing Limited
Published in 1971 by Panther Books Limited
3 Upper James Street, London W1R 4BP

First published by W. H. Allen & Co Limited 1969
Copyright © Paul List Verlag KG, Munich, 1966
Translation copyright © W. H. Allen & Co Limited
and E. P. Dutton & Co Inc 1969
Line drawings by H. Skaruppe, Hamburg
The Quotation from *Half Mile Down* by William Beebe
on page 33 *et seq.* is reprinted by permission of Duell
Sloan & Pearce, affiliate of Meredith Press, copyright
© 1934, 1951 by William Beebe
Made and printed in Great Britain by
C. Nicholls & Company Ltd
The Philips Park Press, Manchester
Set in Monotype Times

Contents

Illustrations

Prologue

The Laboratory of Creation

OUR age is watching a development of incalculable significance: the biologists are discovering "technical inventions" of nature which engineers would have considered Utopian. So it is well worth probing into the huge laboratory in which creation has now been experimenting most successfully for three and a half billion years.

While this book was first being written (1963), more than two thousand biologists and engineers from all over the world met at Dayton, Ohio, to exchange the first results of their joint probe in the realm of nature. In their reports, significantly, one special field was dominant over all the others: research into animal and human senses, and attempts to reproduce artificially the feats, at present beyond our understanding, achieved by the sensory organs.

If we were to get a technological grasp on magnetic and electric senses, heat-registering eyes, butterfly antennae, image-hearing, pattern detectors, electronic instincts, and artificial intelligence, there is no telling what the consequences might be: perhaps a fully automatic car steering through dense traffic without human aid; an artificial dog's nose capable of tracing submerged submarines through the oceans or exposing criminals; a robot which understands speech and can replace the typist in the office; tactile equipment that will forecast earthquakes, and many other things that would change the face of our world.

As for the mysterious realm of the senses, we are discovering things undreamed of before. For the first time we are coming to realize that our "natural" senses are not so natural after all, but rather a kind of magic: the fact, for instance, that our brain composes a visual image from the nerve signals of our eyes is an almost incredible *tour de force*, very rare in the animal kingdom. The world as seen by frogs, bees, and deep-sea fishes is quite different, and it is only now that we are slowly beginning to grasp how that world is made up in each case.

What underlies the phenomenon of pain? How are sense impressions changed by psychological influences? By what technique has nature endowed certain strange animals with "extrasensory perception"? How do migrating birds "navigate" all over the world? Was there ever a "third eye"? All these are questions to which scientists have just begun to find answers.

I hope in this book to offer an overall view, at least in outline, of the great intellectual adventure on the steadily receding frontiers of the unknown. Knowledge of such things should find a place in everyone's general education, of which science is now recognized as an integral part. The knowledge forms a basis for the technological future which is developing so fast and furiously, and it will have a decisive influence on our ideas about the nature of things, on our general philosophy and attitude to life.

Chapter One

The Sense of Vision

The Problems of Seeing

EXPERIMENTS have acquainted us with a paradoxical fact:
man can see "correctly" only because of his imagination. The
human eye, optically speaking, is a piece of bad workmanship.

If the image projected onto the retina by the "camera" –
our eye – were examined by an optician used to high stand-
ards, he would be disgusted. There is more blurring at the
edges than with a cheap pair of child's binoculars; straight
lines look curved, and the outlines fade away under iridescent
haloes. Yet man, who makes such high demands on spectacles, cameras, binoculars, and microscopes, does not notice
anything at all of the shortcomings of his own optical gear:
the nervous system corrects these faults so perfectly that we
perceive a technically flawless image of our surroundings.

This staggering conclusion was reached by Dr. Anton Hajos
of the Institute of Experimental Psychology at Innsbruck
University, and further proof was supplied by an exciting
series of tests. For a number of days or even weeks, Dr. Hajos
and his students wore spectacles with badly distorting prismatic lenses. Here is part of his report:[1]

> While the experiment is going on, the subject is con-
> demned exclusively to a world reshaped by prismatic
> lenses, in which straight lines appear curved, angles are

distorted, and sharp outlines seem fringed with colour.
Objects are not where the subject thinks he sees them, and
they perform ghostly movements as soon as he moves
his head; heavy objects seem to skip about when he
ventures a few steps.

But it does not take long for the grotesque world of the prism
victim to appear normal. Gradually the distortions, coloured
fringes, and "apparitions" dwindle, and after about six days
the subject recovers his impression of a normal, stable, and
optically almost perfect image of his environment. The nervous system has compensated for the conjuring tricks of the
spectacles by its processing of the transmitted image.

This is not all, however. When the subject takes off his
prismatic glasses, the world once more appears to him as if
reflected in comic distorting mirrors – except that now the
straight lines curve the other way, and outlines are blurred by
the opposite colours.

Another experiment carried out by Professor Ivo Kohler,[2]
also at Innsbruck University, yielded even more surprising
results. Test subjects, at the state of seeing coloured fringes,
were taken to a room lit only by sodium vapour light. Pure
sodium light is a monochrome yellow, containing no trace of
other colours; that is why people in such a room normally see
all objects only in shades of yellow. Not so the wearers of
prismatic glasses, who saw colour fringes in normal lighting,
or those still under the prismatic influence. To both these
groups all objects appeared "in glorious Technicolor," which
in fact was not there at all.

"Modern psycho-physics," says Dr. Hajos, "rejects the
idea that the eye – or rather, vision – is an image-producing
optical system." So we can consider it quite miraculous that
our nervous systems manage in the end to synthesize the
defective and distorted information received about our
environment into images tallying with a reality of which the
nervous system seems to have no direct grasp. It is one of the
aims of present research to penetrate more deeply into this
miracle.

For it is by no means certain yet whether at some stage in
our lives, perhaps during the infant's first groping experi-

ments, the brain had to acquire its concept of straight lines, or whether we have an innate feeling for straightness.

The latter is not so impossible as it seems at first sight. Chickens, for instance, have a purely instinctive capacity for recognizing grain shapes. Even if they have never been shown by an "experienced" hen what is edible, from the very first day of their lives they will peck at anything that has the shape of a small grain. In the same way the whole animal world, from the ant to the elephant, has an innate capacity for recognizing shapes and colour patterns, some quite complicated, which are unmistakably characteristic of other members of their species, their mates, rivals, prey, and enemies. I shall have more to say about this later on.

These probings go to the roots of a philosophical theory of perception which I cannot dwell on in this book; my object is merely to provide the scientific basis for further reflections. There is one point, however, which I would like to make at once: no physiological conclusions in favour of any one of the main hypotheses has so far been reached; it seems, rather, as if more than one theory contains elements of truth, and that in the last resort wisdom lies in a synthesis.

Experiments with the Eyeball

Human vision, then, has practically nothing in common with a camera. Further evidence of this can be seen in the results of research into the mechanics of eye movements, carried out by Professor Derek H. Fender[3] at the California Institute of Technology.

Most people are hardly conscious of the fact that only a very small sector in our field of vision can be registered sharply by the eye, a sector no larger than a thumbnail looks when we stretch out our arm. For the visual cells, the rods and cones, are not distributed evenly over the retina: the cones are concentrated, in sufficient density (15,000 cones to the square millimetre) in a tiny spot, the *fovea centralis*. This spot is just large enough for viewing the thumbnail-size sector. Nothing in the surrounding area looks sharp.

I should like to mention here, by the way, that the acuity of vision of an animal equipped with "camera" eyes also depends on the density of the visual cells in its *fovea centralis*, which varies greatly from one animal to another. In the lion the ratio is similar to a man's. According to data supplied by Norman Carr,[4] a lion can still see its prey clearly at a distance of just under a mile. Elephants and rhinoceroses have as few cells in the *fovea centralis* as man has at the edge of his retina. That is why they see their immediate surroundings as blurred, just as things at the edge of our field of vision appear to us. When the distance is over a hundred feet, these animals can hardly make out even objects with quite large outlines.

At the other extreme there are the hawk's proverbially keen eyes. It owes these to the extraordinary density of its cells of vision: its vision equals that of a man with a pair of binoculars that show things eight times their size. But there are other birds too with remarkable vision. Lorus and Margery Milne[5] report that falconers used to take along a shrike in a cage when they went hawking. This little bird was afraid of the hawk, which flew higher than the human eye could see, and would always turn its head in the predator's direction to keep the enemy in sight. So the falconer always knew where his hawk was.

The fact that the centre of sharp vision for the human eye is only of thumb-nail size means that a larger image has to be explored by almost imperceptible, lightning-fast movements of the eyeball. The manner of these exploring movements may supply us with some first insights into image analysis. That is why Professor Fender made the experiment given opposite.

In the same way that some people wear contact lenses instead of spectacles, he fixed glass lenses to the cornea of his test subjects; a little rod with a mirror was fused to the side of the lens. When Fender reflected light onto the mirror, the ray was deflected according to the eyeball's movements. Its route was registered on photographic paper, showing even the most minute movements carried out by the eyeballs, with the assistance of its three pairs of muscles, in order to take in the whole scene.

Oddly enough, the eye is never completely still, even if it is staring fixedly at a given point. It performs involuntary tremor

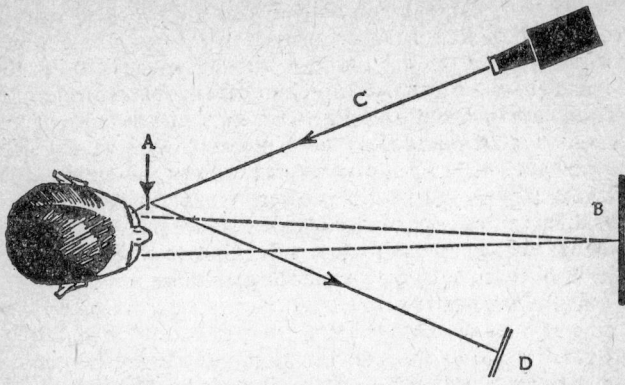

Figure 1. The subject wears a contact lens to which a mirror has been fused on the left eye (A). While the eyes scan the picture (B), the mirror deflects the ray of light (C) and thus registers on the photosensitive paper (D) a graph which records the greatly enlarged movements.

movements, the range of which varies between that needed to view the width of an "i" or an "o" respectively, at normal reading distance.

Figure 2. This image forms on the photo-sensitive paper of Figure 1 when the eye is fixed on one point. Instead of focusing on the small black circle in the middle, the eye ranges over a large area.

This tremor has two advantages for man. One is that it blurs the screen of the picture. Between the retina's individual cells of vision there are optically dead areas which really ought to make our images of our environment look as if each were criss-crossed by a grid of black lines. We should then see everything as though we were sitting too close to a tele-vision screen that, besides its horizontal lines, had vertical ones.

The second advantage can be strikingly demonstrated by stabilizing the image by projecting it from a tiny projector mounted on a contact lens. Every point of our field of vision is now bound to hit the same visual cell all the time. The effect is most peculiar: after a few seconds the image fades and gradually disappears altogether. What remains is a blurred field of grey light, entirely without shape or colour. Even this darkens later on; and as we have lost the sensation of light, the field turns pitch black.

How is this to be explained? Just as a hand gripping an object very rigidly will soon lose its physical impression of the object it is touching, so our cells of vision get used to the constant stimulus and stop sending signals to the brain. Without the tremor of our eyeballs, we should be like the frogs. A frog's eye performs no involuntary tremor; and its field of vision is like a blackboard wiped clean. But anything moving stands out immediately as sharp as a silhouette – an excellent method for the frog to recognize its prey or enemies even if they have taken cover. For man, however, this would be an unimaginable handicap.

But this is not yet a full description of the phenomenon. When the test subject has his image stabilized, so that he loses his vision and sees only a pitch-black wall, an extraordinary thing happens: the image he saw originally suddenly shows up again like an apparition, but only in part. Then it disappears again, and another section of the picture emerges from the darkness. This fades to make room for a third fragment, and so there is a constant alternation – which is even subject to certain rules, as discovered at McGill University in Montreal by the psycho-physiologist Dr. Roy Pritchard.[6]

The break-down of the image into fragments depends on its character and content. If, for instance, it is a woman's head showing in profile, at one time only the face will appear, at another the hair and the back of the head, or the throat and jaw. Each configuration is a meaningful part of the whole. It is the same with words: BEER, for instance, turns alternately into PEER, PEEP, BEE, or BE.

Figure 3. These ghostlike images form in the human eye during the experiment described. From the complete picture (far left of each row) the eye unconsciously picks out the meaningful parts. But it seems also to have a "sense" for abstract geometrical arrangements as the two bottom rows show.

This result would seem to confirm the holistic theory that man takes in only meaningful wholes. But the reverse can also be proved: if the test subject's eye is offered a chequered pattern of squares, individual squares will appear to him, horizontal, vertical, or diagonal, that is, "meaningless" details, geometrical arrangements; man obviously has a sense for these too.

A Computer in the Human Eye

When we are looking for the nerve connections that hold the secret of our receptivity to external impressions, the retina of the eye is the first link in the chain, as Professor Fender discovered in the experiment described above. The first question is: how does the eye manage to bring different distances into focus?

One is at first inclined to assume that it works like a camera with an automatic distance meter: the angular difference between the two visual axes directed at the same point is a measure of the distance, which is transmitted to the muscles adjusting the focal distance of the lenses. This does apply, although with a one-eyed person the lens adjusts itself without the distance having been measured first. So sharp focusing can also take place independently. The focal distance lengthens and shortens all the time. The lens is permanently "on the prowl," and, conversely, also informs the parallax adjustment system about the focal point. The two systems work separately, but exchange their information.

A similar exchange of information takes place between focus control and iris "diaphragm control": as the focal distance of the lens increases, the pupil has to widen in order to keep the brightness of the picture constant.

Are there merely a number of highly efficient tracing and scanning systems and servomechanisms complementary to each other that enable the eye to adjust itself for height, direction, distance, parallax, iris diaphragm, and picture analysis? Or is even more than this involved? Professor Fender had the eye follow an evenly directed movement; that is, a luminous point describing a curve, a series of steps, or a sloping pathway. This showed that, besides having a perfect grasp of the up-and-down rhythm, the eye even anticipates the direction of the movement by about six milliseconds. It "leads the target" in its aim, like an anti-aircraft gun. So the eye must have at its disposal a computer which anticipates the direction of the movement.

In trying to find the seat of this computer, research has hit upon some very strange facts. It cannot be in the human brain; the time for transmitting information between eye and brain is much too long. Light falling on the retina at medium brightness stimulates the nerve cells after a reaction period of 30 milliseconds. The nerve pulses take 5 milliseconds to reach the brain, which needs 100 milliseconds to digest the information. So human perception of anything occurs, in fact, 135 milliseconds after the "event" perceived. Such delay would be intolerable for the mechanics of our eye control – it would mean, for instance, that we could never see an aircraft in sharp focus.

That is why the computer – a mechanical reproduction of this device would have to be as big as a piano – is obviously located in the eye itself, in the millions of nerve cells found in the retina apart from rods and cones. When looking at the retina through a microscope, we notice that it has an astonishing resemblance to the nerve structure of the brain. It was, after all, once part of the brain, which broke away in the course of evolutionary history, as it still does during foetal development. Part of its nerves became readapted to specialize in receiving light stimuli, while millions of others inside it remained normal nerve cells and took over the complicated task of controlling the eye.

The Route of the Nerve Signals

In every human eye the optical image strikes about 130 million visual nerve cells. Every one of these, depending on the intensity of light which hits it, transforms its share of the image into a sequence of electrical discharges – so-called "impulses" or "spikes." The brighter the light, the faster the discharges of the cell. These signals are first passed on by nerve fibres to two other types of nerve in the retina, and it is only from there that they are passed on to the brain. The thick nerve bundle running from the retina to the brain, however, carries only about one million of the nerve fibres. By means of the two other types of nerves mentioned above, nature has invented an "economy circuit," making all but one of 130 conductors redundant, though without involving any loss of quality.

About an inch and a half behind the eye, the two optic nerve bundles from the right and left eye intersect at a point which is called the chiasma. Here the right half of the left eye's bundle crosses over to the right hemisphere of the brain, and the left half of the right eye's bundle to the brain's left hemisphere. Taking into account that the lens of the eye reflects only an upside-down mirror image onto the retina, in the final analysis the brain's left hemisphere receives only impulses from the right half of the image, while its right hemisphere receives only optical impressions of things in the left half of the image. War casualties with one of their visual centres destroyed by a shot in the head (in rare cases such people survived their injuries) see only a half-moon-shaped section of the image.

About an inch behind the chiasma the nerve fibres end in the primary visual centre, and here each optic nerve fibre seems to have its quite specific point of entry. This was shown in an exciting experiment carried out by Professor Jerome Y.

Lettvin[7] of the Massachusetts Institute of Technology, who performed an operation on a frog to remove all the nerve bundles between the eye and the primary visual centre. In contrast to man, a frog has an enviable capacity for regeneration; within a few days new nerve fibres started sprouting

Figure 4. The visual cortex (Z) of the right hemisphere of the brain registers only the left half of a panorama (A-B-C), while the left hemisphere registers only its right half. The explanation for this is given by the course of the visual nerves from the eye through the chiasma (X) and the primary visual centre (Y) to the visual cortex (Z) of the cerebrum.

from the eye towards the primary visual centre, and every one of them struck the exact geometrical spot where the old connection had been. It is obviously very important, therefore, that each optical point of an image also has its connecting nerve point at a geometrically appropriate place in the brain. But at present it is still a mystery to us how the growing nerves find their right way to that body.[8]

The primary visual centre is a relay station. From here new nerve fibres run onto the visual cortex. But it would be completely mistaken to regard the relays in the primary visual centre, and those in the retina too, as simple amplifying stations which pass on the received signal unaltered; this is disproved by the fact that there are 130 million visual nerve cells in th retina, but only one million nerve fibres. The information coming in, then, is also transformed in the relay stations.

To gain a better understanding of these processes, let us take a brief look at the way a nerve cell works.

The outgoing pathway of a nerve cell, the axon, branches out like a river delta in such a way that its ends, with their terminal buttons, either lie directly against the body of the next nerve cell, or they touch one of its incoming pathways (or dendrites). These points of contact between two nerves are called synapses. At the moment that an impulse hits a nerve end, a chemical substance is released and affects the new nerve cell.

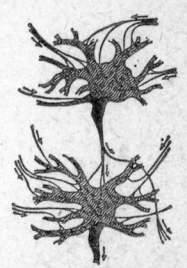

Figure 5. This is how the nerve cells are connected with each other. The axon (black) branches out and touches several succeeding nerve cells with its terminal buttons (black dots) at their centres or dendrites (shaded).

Strictly speaking, there are two different chemical substances and accordingly two different effects. One of these substances excites the new nerve cell; that is, it brings it nearer a state in which the new cell too is capable of sending out an impulse. The other substance inhibits the excitation of the new cell. Whole nerve cells are either excitatory or inhibitory, and it is unusual for synapses *within* a single nerve cell to be mixed. Every nerve end has either one substance or the other and is able only to excite or inhibit.

So at the new nerve cell there is always a kind of contest going on between synapses of excitation and inhibition. This may take quite a dramatic form, for there are sometimes between fifty and one hundred nerve branches ending near a nerve cell that are struggling with each other. There is always a clear-cut decision. If despite the resistance of the inhibiting synapses the "excitators" succeed in exciting the new nerve cell beyond a certain threshold of stimulation, it will fire an electrical impulse through its axon at the next nerve cell, where the whole performance will start afresh. If not, the new nerve cell remains at rest.[9]

In other words, a nerve cell can either bring about or refuse the transmission of incoming signals. Moreover, in a state of increasing excitation it will develop a higher rate of firing. This means that the nerve can add and subtract or multiply and divide. On the other hand, a preceding nerve cell does not have to rely entirely on the decision or "arithmetic" of an individual nerve cell succeeding it. For as a rule branches of the preceding cell's axon also reach other nerve cells, and their decisions, owing to different interconnections, may turn out to be quite different.

How the Image Forms in the Brain

These processes take place in millions of nerve cells simultaneously, all in some way having excitatory or inhibitory effects on each other. When we try to keep this in mind, it gives a very fluid and rather bewildering idea of what is happening each second in our visual nervous system when we look at something.

Nevertheless, American research workers have succeeded in following the route of individual signals from the retina up to the visual cortex and registering the way in which they are transformed. These experiments are really quite breathtaking.

It is impossible, of course, to discover the path of individual nerve fibres by optical means, even with the help of an optical or electron microscope. Purely from the outside, moreover, an excitatory synapse looks no different from an inhibitory one; and besides, every nerve cell is at a different stage of excitation every tenth of a second, so that static observation will do very little good. Consequently another method must be used.

This consists in electrical recording from individual nerve fibres with simultaneous pinpointed exposure to light of the corresponding sensory cells in the retina. It often takes hours to strike the right point of the retina with the microscopically small point of light. You can tell a hit by a sudden volley of discharge from the tapped fibre. These experiments were not made on human beings, of course, but on anaesthetized cats.

Just a word on the recording technique: little glass tubes filled with a fluid which conducts electric current have proved useful as electrodes. These are so thin that thousands of them might just about fill the point of a pin; they are hardly perceptible to the naked eye. That is why they have to be inserted into the nerve cell, or brought up very close against a nerve fibre, by means of a micromanipulator under the control of a stereomicroscope. Then the electric nerve impulses are conducted into an amplifier and made visible on a sort of television screen, a so-called cathode ray oscilloscope, or recorded as a graph.

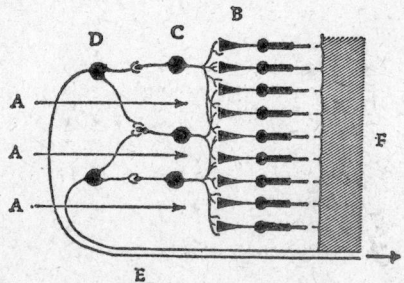

Figure 6. The two first sorting stations in the optical nervous system. The light rays (A) come from the left and excite the visual cells (B). Their impulses are collected and sorted first by the bipolar cells (C) and then by the ganglion cells (D). From here the optic nerves (E) run without further interruption through the rear wall (F) of the eyeball to the brain.

This is the route of the pictorial information: from a number of visual receptor cells fibres lead to the first "sorting office", the bipolar cell, which is also embedded in the retina – while every one of these visual receptor cells is connected to several other bipolar cells. Similarly, the bipolar cells are connected to the second station; that is, to nerve cells called ganglion cells which also lie within the retina.

To watch what was happening at these first two stations, Professor Stephen W. Kuffler[10] tapped the ganglion cells one by one at The Johns Hopkins Hospital, Baltimore, Maryland, and arrived at the following result: each ganglion cell is influenced by all visual nerve cells within a sharply defined circular section of the retina. Any light stimulus striking within the range of the circle excites the neuron. There is a slightly larger ring-shaped section surrounding this smallish excitation

centre. If a light stimulus strikes the outer ring, it has an inhibitory influence on the same neuron. A spot of light on the
excitation centre and an equally strong one on the inhibiting
ring roughly cancel each other out. This type of double circle
is called an excitation-centre field.

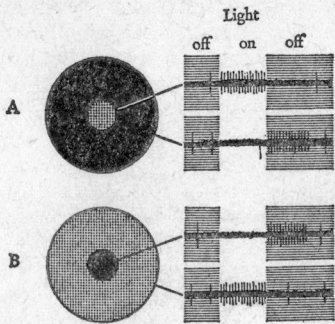

*Figure 7. This is how the signals are collected and sorted. All the
visual cells on the retina within the circular fields affect a single
ganglion cell. If a visual cell is in the dotted areas, on exposure to light
it stimulates the ganglion cell to send out impulses, so-called 'spikes'.
If it is in the black area, it suppresses the 'broadcasting' activity of the
ganglion cell until the light goes out. There are only two types of
circular fields: that of the excitatory centre (A) and that of the
inhibitory centre (B).*

But besides this the retina also contains a second, reverse
type of ganglion cell, the inhibitory centre. Every light
stimulus that strikes the centre of the circle decreases the
nerve cell's firing speed, while every light stimulus striking the
surrounding ring increases that speed.

The whole image that we see is divided into millions of such
reactive fields. The analogy of a mosaic does not apply, for
the individual regions overlap. The size of these so-called
receptor fields varies a great deal depending on their position
in the retina.

The significance of this division of the image is difficult to
grasp in purely concrete terms. We can only imagine that it
may serve to heighten the contrast in the fovea and also the

effect of dim twilight at the edge of the retina. All this, however, has very little to do with the electrical transmission of a picture as we know it from television techniques.

The situation in the primary visual centre is very similar. This was investigated at Harvard Medical School by Professor David H. Hubel,[11] the neuropsychologist. He experimented also with anaesthetized cats. Again many fibres from the ganglion cells converge in each nerve cell, while every ganglion cell is connected to several nerve cells in the primary visual centre. Again every nerve cell of the retina has a ring-shaped excitation or inhibition centre surrounded by a ring with the opposite influence. The arrangement, however, is a little different.

But things are completely different as soon as these nerve fibres from the primary visual centre reach the visual cortex of the cerebrum.

First I should perhaps give a brief survey of the cerebrum's structure. If the human brain's cortex, which is convoluted like the inside of a walnut, were flattened out, it would be 20 square feet in area. At a depth of between 0 and 2·5 millimetres it harbours no less than 14 billion nerve cells, in which take place the miracles of sensory perception, feelings, thoughts, and creative powers. These nerve cells are the seat of intelligence, personality, and character.

Strictly speaking, there are seven different types of nerve cells arranged in seven layers. The thickness of each layer changes abruptly at every functional boundary, for example, between the visual and auditory areas.

The millions of nerve fibres coming from the primary visual centre join the nerve cells in the fourth-from-top layer of the visual cortex. From here communications with all the other layers branch out, nearly all running perpendicular to the brain's surface. From these layers, particularly the third and fifth, axons pass into deeper regions of the brain, about which we so far know nothing.

Hubel recorded from a few hundred of these nerve cells one by one, and found every one of them connected to entirely different receptor fields in the retina. Not a trace now of tiny centres with concentric rings; instead there is a retinal area about a millimetre square that influences an individual

nerve cell of the visual cortex. This is amazing, for such a
large part of the retina accommodates a section of the image
equivalent to thirty disks the size of the full moon. There are
between ten and twenty excitatory points and about as many
inhibitory points in this region of the retina, sometimes with
fairly large spaces in between, arranged in such a manner that
the two types are divided by perfectly straight borderlines.
There are also so-called slit regions, in which the excitatory
points are strung along the middle like beads on a string
surrounded on both sides by inhibitory points.

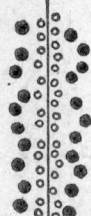 *Figure 8. Detector field for vertical, straight light lines.
This section of the retina of our eye is really about one
millimetre long. All the visual nerves in the white fields
excite a single nerve cell on entering the visual cortex
of the cerebrum. All the visual nerves in the blackfields
on exposure to light have an inhibitory effect on the
same brain cell. The vertical line shows the direction of
maximum excitation.*

If a strip of light one millimetre long hits all excitatory
points directly, discharges from the nerve cell in the visual
cortex are fastest. But if the strip is turned by only a few
degrees, like a propeller, the speed of these discharges at once
goes down rapidly, ceasing entirely at an angle of 90°. This
means that the nerve cell is a detector of straight light lines,
but only if they run in a particular direction.

There are countless other nerve cells that are in principle of
the same excitatory type. But with *them* the direction of the
"slits" is turned by any angle between 0 and 180°. The brain
can thus perceive straight light lines in any direction.

This fact points to an innate sense for straight lines. Remembering the experiments with distorting spectacles described at the beginning of this chapter, we are faced again by a
theoretical problem of perception: what happens in the
nervous system when the experimental subject gradually
starts seeing the curved lines as straight again? Do the contacts – that is, synapses – then alter throughout the whole
optic nerve system? We do not know. In any case there is
close co-operation between innate and acquired knowledge.

Besides the type of nerves in the visual cortex mentioned above, there are others for the perception of straight dark lines, and others again for the detection of edges (excitation on the left of a straight borderline, inhibition on the right of it, or vice versa). A further type does not react to static lines but does to moving ones, while each nerve cell is responsible for only one specific direction of movement.

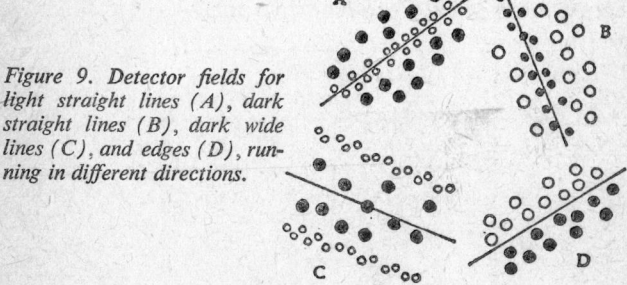

Figure 9. Detector fields for light straight lines (A), dark straight lines (B), dark wide lines (C), and edges (D), running in different directions.

The break-down of the image into a vast number of small edges, light and dark lines, explains the tremendous number of nerve cells in the visual cortex of the cerebrum. For every section of the retina, every type of line (slit, strip, edge), every position of the line and direction of movement, there is a set of nerve cells which always responds and passes on impulses. Even when the eye sees no more than a propeller rotating against a white background, the number of brain cells excited by it is unimaginably great.

After this statement one might imagine a bewildering confusion of all these nerve cells in our brain. Professor Hubel's discovery that in fact there is exact order is all the more surprising. It is an order, however, that is not discernible by the look of things under a microscope but only by their electrical behaviour.

According to these observations, the visual cortex of our cerebrum is divided like a honeycomb into innumerable columnar segments. Each column has a diameter of about half a millimetre and consists of thousands of nerve cells. The

common denominator of all these nerve cells is the direction
of their stimulation fields in the retina, which may be shaped
like slits, strips, or edges. All pictorial outlines from a section
of the retina, which run at an angle of 30°, are registered in
this one column. The neighbouring columns, however, only
respond to lines and edges running at an angle of 15°, or 31°,
33°, and 35° respectively; while *their* neighbours deal with
other angles. There is no symmetry between the arrangement
of the excitatory points within the retina and that of the
nerve cells they stimulate within the column.

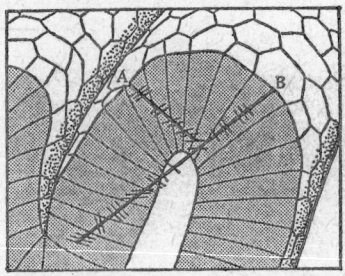

*Figure 10. The "honeycomb" of the visual cortex. Each nerve column
responds to lines in one direction only from a small section of image on
the retina. The next columns deal with lines in other directions. This
was established by gradually shifting two microelectrodes (A and B),
which monitored the signalling activity of the nerves, while a slowly
rotati   of light was brought to bear on the retina.*

Fascinating as the details of this research work may be,
they make the process of seeing even more mysterious than it
seemed before. The breaking up of the black-and-white
picture into innumerable lines; the jumbling of spatial relations; the simultaneous coded messages flashing through
millions of channels; the nerve fibres growing in the right
direction to find their point of contact; the nerve circuits
beyond human grasp and yet so purposeful in their multifor-
mity; the susceptibility of these circuits to processes of learning; the co-ordination of sensory stimuli received with the
tremor and probing movements of the eyeball – all this adds
up to a great miracle of creation. It is only in our day, through

the advances in scientific research, that we are beginning to get an inkling of its full greatness.

The Electronic Frog's Eye

In the United States, the country where "the future is already here,"* all talk about remote-control cars with steering wheel, acceleration, and brakes to be controlled by a buried high-frequency cable, stopped in 1963. This project, which would have degraded cars of the future to a sort of miniature trolley, was scrapped before there was any chance of realizing it. It has been replaced, in our age of hectic technological development, by an even more astonishing one: the electronic frog's eye.

The American scientists Lettvin, Maturana, McCulloch, and Pitts[12,13] have gleaned from the eye of a perfectly ordinary frog an optical and mathematical principle on the basis of which they are to develop a fully automatic device, an electronic driver, which will be able to steer a car safely through heavy traffic.

This sounds Utopian, yet is quite possible, as nature proves by the frog's example. But one has to realize that here nerve circuits of an entirely different kind respond to the optically received picture, to form impressions that have hardly anything in common with our image of the world, though they are very practical for the frog!

Figure 11. Schematic circuit for an 'artificial nerve cell'. Cell (A) sends an impulse through axon (B) only when it is stimulated by excitatory centre (C). If a signal from inhibitory circuit (D) arrives at the same time, the artificial nerve cell shows no reaction.

Professor Heinz von Foerster[14] of the University of Illinois has explained the easiest way for man to penetrate into the

*'The Future Is Already Here (Die Zukunft hat schon begonnen) is the title of a successful book by Robert Jungk.

strange world of images as seen by frogs and many other animals: electronically, it is child's play to create artificial nerve cells, rough simplifications of the natural structures described above (p. 19). With this alone, amazing combinations of circuits can be formed. Let us start quite simply: out of a row of photo-electric cells each is connected to its artificial nerve cell by two excitatory conductors, and by one inhibitory conductor each to the two neighbouring nerve cells. If the light exposure of all the photo cells is equal, the nerves discharge no impulses, because every nerve is twice excited and twice inhibited. So this eye-and-nerve combination does not "see" anything.

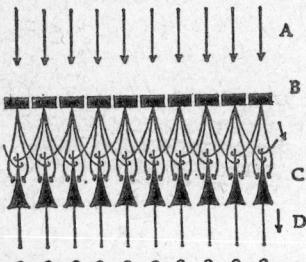

Figure 12. This combination of artificial nerve cells 'sees' nothing if light (A) falls equally on all photo cells (B). Since each artificial nerve cell (C) is twice excited and twice inhibited the cells discharge no impulses (D).

But directly an object enters the field of vision, its outlines are immediately detected, as the nerve cells beneath its edge are now excited twice and inhibited only once, which means that they pass on an impulse. The following experiment can be made: the photo-cell "retina" may be enlarged and refined to any extent. We then scatter a handful of peas in front of it. An electronic device, which tots up all the information received and divides it by two, then reports the number of peas almost at once.

This very simple counting machine, therefore, has a faculty the human nervous system does not possess: instead of counting by consecutive numbers, as we do, it takes in everything at a glance. It sees a quantity as we see certain electromagnetic vibrations – as a concept in its own right. If, for instance, the retina's visual cells are struck by light with a frequency of 385 million kilocycles per second, we do not have to count them to know that it is red light. This has already been done for us by a nerve circuit, which simply informs our consciousness of its result: red! In a completely analogous way the device described above immediately registers a "seven-ness," a "sixty-eight-ness," or a "27,695-ness."

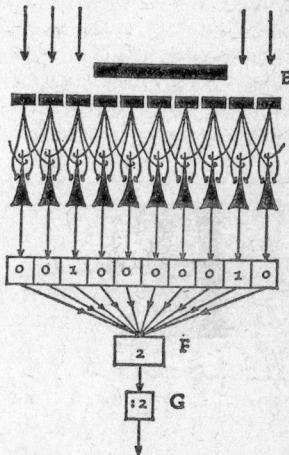

Figure 13. As soon as the shadow of an object (E) falls on a photocell, the artificial nerve cells at the edge send out an impulse, since they are now twice excited and only once inhibited. An electronic device (F) adds up the result, another (G) divides it by 2. In this way the counting machine grasps like lightning the number of objects to be counted.

So we have here not only a prototype of technical counting machines, but also an impressive example of the construction of an entirely new sense, beyond our imagination, an "absolute number sense," based solely on an unusual nerve circuit.

It is an exciting prospect to expand this fairly simple principle, perhaps by adding a second dimension. One may also get the artificial eye to revolve. This would make the counting machine into a device capable of tracing outlines and detecting forms by mechanical means. Such devices can distinguish between straight lines and curves, and detect holes, corners, and other characteristic shapes. In modern research they serve as a foundation for further refinements which will enable them to read figures and letters, evaluate photos and X rays, detect cancer cells in microscopic preparations.

But what has all this to do with the frog's eye? A great

deal. For these amphibians have similar circuits for recognizing patterns, and so have octopuses and insects. Only we have to break away from the untenable idea that all animal eyes are designed to provide the animal with photographic, true-to-life images.

As stated above (p. 16), the image rendered by a frog's eye is like a blackboard wiped clean, on which only the moving object appears; that is, something the frog either fears or wants. But even these objects are not seen in their natural shape. When Professor Lettvin investigated the frog's visual nervous system by microelectrodes in the same way that Professor Hubel had done with the cat, his findings were very strange.

The basic elements into which the optical image is broken up by the frog's nervous system are different in kind and far less universal. The frog's brain is informed about only four things: straight edges, moving convex forms, change of contrast, and rapid darkening.

But this information does not even comprise every direction of movement (as it does with the human eye). If a fly is heading for the frog so that he might catch it, his brain registers it. But if it is moving away or flying so that it remains out of his reach, it simply makes no "sense" to the frog. He *doesn't see* anything that does not concern him directly – a strikingly practical principle of restriction to what is absolutely essential!

So one can imagine that the frog likewise does not see blades of grass moving in the wind, because there is no contrast. But if a stork approaches, a series of impulses, only from the nerve cells which react to moving convex forms, and change of contrast, are obviously enough to make the frog jump back hastily into his pond. So abstract is the world the frog lives in, and so impossible for us to visualize.

Two American biologists, M. B. Herscher and T. P. Kelley,[15] suggest that we might in a similar way abstract the image of our traffic-congested streets and split it up into characteristic features indicating the end of a road or people and things we must give way to. Traffic signals would have to be incorporated into the scheme, and so would a steering programme to be fed by the car-owner into the electronic device before the start of a journey. Our automatic driver would disregard as irrelevant all other activity taking place on the roads.

But before this goal can be reached, we shall have to make sure that man scores considerably better than nature for reliability; otherwise the "electronic frog" will after all be eaten one day by the stork.

Creatures that See only Light Patterns

LUMINOUS DEEP-SEA FISH

"As 600 feet came and passed, I saw flashes of light in the distance. They were animal illuminations, resembling in appearance the first stars on a warm, clear summer's evening. I looked out into a world of inky blueness where constellations formed and reformed and passed without ceasing." This is how William Beebe,[16] the American deep-sea explorer, describes his first descent into lower depths in a bathysphere. Here are some further extracts from his reports:

At 2,000 feet I made careful count and found that there were never less than ten or more lights – pale yellow and pale bluish – in sight at any one time. Fifty feet below I saw another pyrotechnic network, this time, at a conservative estimate, covering an extent of two by three feet. I could trace mesh after mesh in the darkness, but could not even hazard a guess at the cause. It must be some invertebrate form of life, but so delicate and evanescent that its abyssal form is quite lost if we ever take it in our nets. Another hundred feet and Mr. Barton [his companion] saw two lights blinking on and off, obviously under control of the fish. . . .

From 2,050 to 2,150 feet I saw relatively few illuminated organisms, but later, at 2,200 feet, the lights were bewildering. . . . Pteropods were close at hand and a host of unidentifiable organisms. I would focus on some creature, and just as its outlines began to be distinct on my retina, some brilliant, animated comet or constella-

tion would rush across the small arc of my submarine
heaven and every sense would be distracted, and my
eyes would involuntarily shift to this new wonder.

. . . the jet blackness of the water was broken only by
sparks and flashes and steadily glowing lamps of appreci-
able diameter, varied in colour and of infinite variety as
regards size and juxtaposition . . . Now and then, when
lights were thickest, and the watery space before me
seemed teeming with life, my eyes peered into the distance beyond them, and I thought of the lightless
creatures for ever invisible to me, those with eyes
which depended for guidance through life upon the
glow from the lamps of other organisms, and, strangest
of all, the inhabitants of the deeper part of the ocean,
those blind from birth to death, whose sole assistance,
to food, to mates, and from enemies, were cunning
sense organs in the skin, or long, tendril-like rays of
their fins.

According to Beebe, the outlines and bodily shapes of
luminous deep-sea fish may be discerned, from very close
quarters, in the faint gleam of the fishes' own light; but at a
distance of a few yards it is practically only the "constellation" that can be seen. Just as a captain at night can tell only
by her lights whether the ship that passes him in the open sea
is a big freighter, a small motor launch, a salvage vessel, or a
pilot-boat, deep-sea fishes also distinguish friend and foe
merely by the light and colour pattern of the lamps which
every luminous fish exhibits in a characteristic way. But this
code of recognition is much more complicated with fish
than with shipping.

Some sea-dragons (*Melanostomiatids*), for instance, look like
lighted passenger ships. *Bathysphaera intacta* (the Untouchable
Bathysphere Fish), as Beebe named him, has a row of twenty
pale blue "portholes" running along either side. There are
tentacles, about three feet long, hanging from its "bow"
and "stern", each with a luminous bait being dragged along
at the bottom end – the upper tentacle reddish, the lower
one blue. Although this light display looks peaceful enough,
its exhibitor is one of the most voracious of deep-sea pre-

dators. His fangs, which are illuminated from inside, seize on any creature that unwarily approaches the baits.

It is just the other way round with the deep-sea hatchetfish of the genus *Argyropelecus*. Their phosphorescent beams form a ghostly pattern reminiscent of the teeth of a human

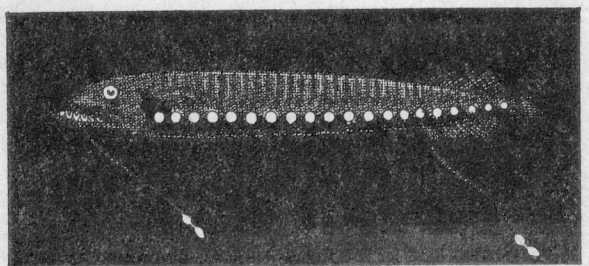

Figure 14. The sea-dragon.

skull. Yet these creatures are harmless plankton-eaters. It has not yet been discovered, however, whether they can frighten away enemies by their fierce appearance.

Figure 15. Deep-sea hatchetfish in the beam of a searchlight. What stands out in the darkness is not the shape of their bodies but only their light organs, reminiscent of the teeth of a human skull

The Five-lined Constellation Fish (*Bathysidus penta-grammus*) – again Beebe's name – is the aristocrat of the deep sea. "Along the sides of the body were five unbeliev-ably beautiful lines of light, one equatorial, with two curved ones above and two below. Each line was composed of a series of large, pale yellow lights, and every one of these was surrounded by a semi circle of very small, but intensely

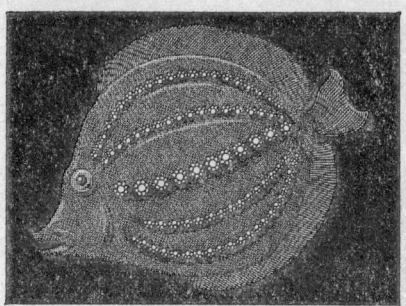

Figure 16. The Five-lined Constellation Fish.

purple photophores. . . . In my memory it will live throughout the rest of my life as one of the loveliest things I have ever seen."

The number of light uniforms is immense. In the family of lantern-fish alone there are about 150 different species to be distinguished by the number and arrangement of their luminous organs.[17] Two genera, for instance, vary only through differences in their light equipment just in front of the tail fin. If the fish have two luminous organs here, they belong to the genus *Myctophum*, but if they have between three and six, they are members of the genus *Lampanyctus*. Differences between the separate species are even smaller, yet the fish recognize them exactly.

Even males and females differ in their "light display" as male and female birds differ in their plumage. A lantern-fish frequent in the Atlantic and Mediterranean is *Mycto-phum punctatum*. The sides of his body are resplendent in the glow of bright "portholes." The females also carry three to five luminous plaques on the bottom side, the males one to three light points on the top side of the root of their tail.

In other words, the vision of those deep-sea fishes living in eternal darkness is confined exclusively to the registering and interpreting of abstract but precisely characteristic light patterns. While the frog's visual nervous system first abstracts typical characteristics of food or danger from the

Figure 17. A deep-sea fish that "angles" for other fish with a lantern

image of his surroundings, the field of vision in the depths of the ocean consists almost entirely of abstract flashing, twinkling, and colourful "neon lights."

There are, however, voracious predator fishes that misuse their lamps to attract prey. With the deep-sea angler fish the first barbule of the back fin has developed into a fishing rod of varying length, with a small luminous balloon hanging at its far end. This bait looks like a luminous marine worm, and is brought forward to dangle in front of the fish's wide-open mouth, which is spiked with huge teeth sharp as needles.

With *Gigantactis macronema*,[18] the fishing rod is four times as long as the whole fish. By means of his very pressure-sensitive lateral-line organ, the fish can locate any approaching creature. At the right moment he will withdraw the rod and pounce on the inquisitive intruder. Things are easier for other species with shorter rods: they suck in the prey swimming around just in front of their open mouths.

The methods of deep-sea barbel fish or stomiatoids are even odder. Instead of rods they have "beards" with a luminous organ hanging at the end which also serves as a bait. The fish of one species measure eleven inches, but their "beards" are over three feet long. Nerves in the "hair" of their "beards" signal the approach of prey. Beebe says of the experiments he carried out in an aquarium with these lantern-carriers: "Even the minutest movement of the water near his beard produced the utmost excitement in the fish. He became savage and snapped, trying all the time to reach the source of the disturbance and to bite."

Chauliodus, a species of the viper-fish family, also a deep-sea denizen (living at depths between 1,500 and 7,000 feet), does not need to locate its prey by pressure waves. These fish are expert illusionists which illuminate the inside of their mouths splendidly with 350 dots of light. Small fishes and crustaceans swim cheerfully into this lethal splendour so that the viper-fish has only to do a little chewing now and then.

Glaring flashes of light serve again a different purpose. Some members of the lantern-fish family (*Myctophidae*) have developed them as defensive weapons. They use them at moments of danger to dazzle the aggressor so much that he leaves them alone. When Beebe suddenly put his watch's luminous dial in front of the head of a lantern-fish (*Myctophum affine*), the frightened creature at once reacted with a series of flashes.

In his bathysphere Beebe had the following experience:

I watched one gorgeous light as big as a ten-cent piece coming steadily towards me, until, without the slightest warning, it seemed to explode, so that I jerked my head backward away from the window. What happened was that the organism had struck against the outer surface of the glass and was stimulated to a hundred brilliant points instead of one. Instead of all these vanishing as does correspondingly excited phosphorescence at the surface, every light persisted strongly, as the creature writhed and twisted to the left, still glowing, and vanished without my being able to tell even its phylum.

As later transpired, it had been a luminous deep-sea shrimp of the genus *Acanthephyra*. In a moment of great excitement it had released a veritable rain of sparks from its arsenal, thus enveloping its enemy in a blinding, confusing sea of light.

Effective though the light-flash is as a defence mechanism, it involves a difficult problem for the user: how can he avoid blinding himself? The solution nature has found for this problem is amazing: with deep-sea hatchet-fish, barbel fish, and others, a small reflector is placed near the eye in such a way that the light from outside is reflected directly into the creature's own eye. The eyes thus become accustomed to the mass of light that is to follow.

But this is far from exhausting all the uses of animal luminescence in the deep sea. In 1959 H. W. Lissmann, the English marine biologist, put forward the theory that the flashing of their lamps served for the fishes of the depths as a means of communication. Perhaps besides characteristic arrangement and colours of the light patterns, there is here an equally characteristic signal code, such as we are familiar with in lighthouses. Perhaps the phenomenon even represents a kind of message in Morse whereby the creatures of eternal night can summon mates, give warning to rivals, and, possibly, "converse" about other things as well. It is an interesting subject still awaiting detailed research.

The position seems to be similar with the luminous shrimp (*Sergestes prehensilis*).[19] The Japanese biologist Terao found that this crustacean's body is bespangled with over 150 dots of light, which can all be switched on and off at lightning speed. Within a second or two, green and yellow light patterns will flit from head to tail in quick succession, like moving neon lights in the centre of a big city at night.

The most uncanny creatures are the luminous squid. Some of these amazingly intelligent creatures, which often hunt in fast-swimming schools, have highly developed luminous organs with lenses, concave mirrors, diaphragms, and shutters. With the genus *Desmoteuthis*[20] one may even, without exaggeration, speak of searchlights. These are moved about by muscles and cast their light backwards and downwards; that is, in the direction the torpedo-shaped body is moving. They light up its route, track down prey, and blind enemies.

Strangely enough, there are two fundamentally different searchlight structures with these creatures: open and closed luminous organs.

Those with open luminous organs, mostly the inhabitants of shallower seas, cannot make their own light. They depend on the assistance of luminous bacteria. In little sacs under their skin or in a ball of tiny tubes they collect these light-givers from the sea-water. To keep their light in the bag, they offer these microbes an attractive culture medium. Five different species of luminous bacteria have been counted in the common squid (*Sepia officinalis*),[21] all of which had settled in separate luminous organs, strictly divided according to their species. Apparently the composition of the culture medium is decisive in whether the bacteria will stay, or leave again. There is a thin tube always open, allowing the guests to come and go as they please.

However, if the squid is in danger, the bacteria may find themselves evicted willy-nilly. In the dark of the deep it would be no use for the squid to emit ink when trying to elude enemies. That is why one of the protective measures taken by *Sepiola ligulata*, *Rondeletiola minor*, and *Hetero-teuthis* is to emit luminous clouds by pressure on their "light tubes" and to confuse enemies by blinding them for three to five minutes.

Things are quite different for creatures with closed luminous organs, mostly inhabitants of the deep sea. There is no need for them first to steal their light; they have glands in which they produce a luminous liquid. Also, like theatre spotlights, they can conjure up the most varied colours. Dr. Hans-Eckhard Gruner[22] writes: "By different devices like coloured filters (skin pigments) and glittering mirrors, all sorts of colour shades may appear in one particular creature. It may produce red, blue, green, or white light. Its luminous organs, when active, will then sparkle in these colours like precious stones."

The variety of forms and colours is inexhaustible. The "infernal vampire squid" (*Vampyroteuthis infernalis*),[23] which has its tentacles connected by membranes to form an umbrella-shaped trap, carries at the end of its body two reflectors on stalks an inch and a quarter long. If it wants to extinguish these, it retracts them into pockets, which it

Figure 18. The infernal vampire squid.

closes with a lid. Since, in contrast to glow-worms, squid cannot switch off their lamps, they have to work with black-out curtains. These exist in several varieties: sometimes an end of the ink-bag or a sort of blind is used. Both are inserted between the source of light and the lens.

The most beautiful light in the depths of the South Atlantic is given out by a creature called *Lycoteuthis diadema*, that is, "diademed squid." It was discovered by the leader of the famous German *Valdivia* expedition, Professor Carl Chun,[24] who gave the following description of it:

> *Lycoteuthis diadema* is endowed with 24 luminous organs, which show a peculiar grouping. There are two on each of the two big tentacles. The lower rim of both eyes is edged by five each, and the rest are distributed over the body in a specific arrangement: on the underside the two forward lights edge the vent, while of the five centre ones the outermost are situated at the base of the gills, and the rest are at the hind end. Whatever brilliant colours other deep-sea creatures may offer us, they come nowhere near the splendour of these lights. The body looked as if it were trimmed with a tiara of brightly coloured precious stones: the central one of the eye-lights had an ultramarine sparkle, and the ones at the sides shone like mother-of-pearl. Of the ventral lights the ones in front had a ruby-red gleam, while the hind ones were snow-white, except for the central one, which was sky-blue. It was magnificent!

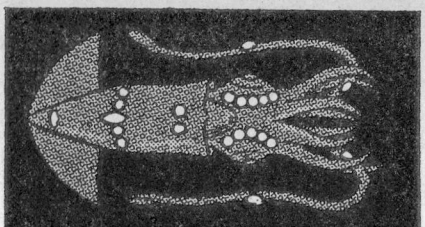

Figure 19. The diademed squid.

LUMINOUS UNICELLULARS

On 27 September 1959, tanker captain W. Rutherford[25, 26]
had an uncanny experience crossing the Java Sea between
the islands of Java and Borneo. The report he gave on it
sounds like a typical old salt's tale:

> Between 2350 ship's time on 27th September and 0010
> on 28th September, 1959 . . . slight sea, very dark and
> clear. . . . The first indication of anything unusual was
> the appearance of white caps on the sea here and there,
> which made me think that the wind had freshened, but
> I could feel that this was not so. Then flashing beams
> appeared over the water, which made the officer on watch
> think that (some nearby) fishing boats were using
> powerful flashlights. These beams of light became more
> intense and appeared absolutely parallel, about 8 ft wide,
> and could be seen coming from right ahead at about
> $\frac{1}{2}$ sec intervals. At the time I thought I could hear a
> swish as they passed, but decided that this was imagina-
> tion. . . . It was like the pedestrian's angle of a huge
> zebra crossing passing under him whilst he is standing
> still. When this part of the phenomenon was at its
> height it looked as if huge seas were dashing towards
> the vessel, and the sea surface appeared to be boiling,
> but it was more or less normal around a fishing vessel
> which we passed fairly close. . . . The character of the
> flashes changed and took on the appearance of beams
> from a lighthouse situated about two miles on the
> starboard bow, or as if the centre of a giant wheel was

somewhere on the starboard bow with the beams as its spokes. As the beams from the wheel on the starboard bow weakened, the same pattern appeared on the port bow at the same distance and regularity. The wheel on the starboard bow revolved anticlockwise and the one on the port bow revolved clockwise . . .

The next change was that the beams appeared to be travelling in the exact course of the ship, like a following sea . . . "chasing" us. . . .

Presently all the beams gradually ceased and the surface of the sea could be seen again. At that time for about two min, as far as the eye could see, there were rings of light about two ft in diameter and six ft apart in the sea, flashing in and out with rhythm. The flashing reminded me of a treeful of glow-worms.

In 1960 Professor Kurt Kalle,[27] former marine chemist of the German Hydrographic Institute in Hamburg, examined seventy similar reports by captains. Fifty-one of them tallied in the essentials. The remaining nineteen sketched a picture as of a bombardment of light balls from the depths. Small luminous balls, about three feet in diameter, emerge from the depths and on the surface burst noiselessly to form glaring light discs about three hundred feet in diameter, which then quickly fade out.

How is this uncanny apparition of the tropical seas to be explained? The light is certainly produced by the small unicellular luminous organisms that populate the sea in their thousands of millions and produce its phosphorescence.

"Scintillating Nightlight" (*Noctiluca scintillans*) is the name of this small dinoflagellate, between half a millimetre and two millimetres in size, which has the shape of a peach with

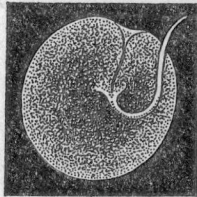

Figure 20. The "Scintillating Nightlight," which causes sea phosphorescence.

a little whip. At certain times of the year, and under certain
weather conditions, directly it is subjected to extra stimulation by pressure waves, hundreds of tiny light dots flare up
inside its transparent body, producing a phosphorescent
gleam.

The pressure waves releasing this phosphorescence may
be produced by ships or dolphins, which will then appear to
drag along luminous stripes behind them in the dark; also
the surf on a beach may look like a wall of fire. When the
sea is rough or in torrential rain, the whole sea glows as far
as the eye can see.

But the light-releasing pressure waves may also hail from
a seaquake. Kurt Kalle arrived at this conclusion because
of the fact that the luminous giant wheels have so far been
observed only in volcanically active sea areas. The phenomenon described by Captain Rutherford would accordingly
have been seaquake waves which had become visible.

Professor Kalle gives the following detailed explanation
for the phenomenon. Above the deep sea, shock waves
emanating from a point source produce the image of light
balls shooting up. Their horizontal bursting asunder on the
surface of the sea would agree with the radial expansion of
the elastic waves. Above the continental shelf, however, the
part of the shock wave reflected downward is reflected once
more by the bottom of the sea, reaching the surface again
practically without any loss of energy. The primary and
secondary quake waves overlap, producing interference
effects. The luminescent organisms then indicate the position
of areas at the same stage (including amplification), while
the interspersing dark sections indicate the coincidence of
opposite phases and the cancellation of the pressure stimulus.

Kurt Kalle has been able to consolidate this theory by
interference models. A complete proof, however – by seismic
readings, for example – has so far not been obtained.

LUMINOUS INSECTS

In the Blue Mountains of the island of Jamaica a strange
phenomenon of nature can be watched at night. As if lit by

a ghostly hand, "Christmas trees" will suddenly flare up here, there, and everywhere in the darkness of the desolate landscape.[28]

A closer approach will reveal the trees, surprisingly enough, to be palm-trees, with blossoms which are flaring like seething fires: hundreds and thousands of small luminous beetles have assembled here. But in contrast to the glow-worms of Europe, these creatures do not glimmer permanently. Rather, there is a great confusion of flashes, every insect flashing twice a second. An observer will thus get a most eerie general impression, as of so many will-o'-the-wisps.

Even if this luminous splendour is swamped by torrential rain, or if the trees are shaken by a storm, the insects (*Photinus pallens*)[29] still go on shining with undiminished brightness; they can't really help themselves! For the party with its gay lights is nothing but a great marriage market. In Europe it is up to every individual glow-worm to attract a male with its lamp. In Jamaica thousands of brides and bridegrooms pool their lights in a collective courting effort. The trees they illuminate shine over a distance in excess of half a mile, attracting more and more of the luminous insects all the time. If you hang up a bright flash-lamp in a palm-tree there in the evening, you may be sure that the first luminous beetles will soon arrive, and that after about an hour the tree will be a sea of flames in which the creatures will be mating.

This courtship by flash-lamp is much more difficult with the closely related black firefly (*Photinus pyralis*), which lives in the southern United States. On his courting flights the male sends out a flash lasting $0 \cdot 06$ seconds every $5 \cdot 7$ seconds.[30] The females, meanwhile, are sitting in the grass, entirely blacked out. But as soon as a flashing male approaches to a distance of between 10 and 14 feet, the female responds exactly $2 \cdot 1$ seconds after each flash, blinking in her turn. Now the suitor immediately makes for her light, and after signals have been exchanged five times, or ten at most, he will have achieved his aim.

Professor Buck, who discovered this phenomenon, has also succeeded in imitating the female with a flash-lamp – though he had to work with tremendous accuracy to press the button exactly 2·1 seconds after the flash of a male that happened to be approaching. If he blinked only a fraction of a second too early or too late, the beetle took no notice of him and simply flew on. Only if he knew the "password," flashing exactly at the interval prescribed, would the insect land on his hand. This is an impressive demonstration of the highly specific signal code which prevents the male when mating from approaching females of other luminous species in the dark.

Luminous beetles of the genus *Colophotia* in South-east Asia have developed an unequalled imitation of neon lights. During the day these insects stay in the Burmese jungle. But in the evening the males fly out to some special mangrove trees along the river-banks, which for reasons unknown they have chosen as their haunts. Here they now start, just like their relatives in Jamaica, to blink twice a second. But in contrast to their Jamaican cousins, the Asian genus all blink at the same time, in chorus as it were. Just imagine a luminous beetle sitting on almost every leaf. This makes thousands of them in a single tree; yet they all, without exception, blink as exactly as if they had been turned on by someone pressing a button. What is more, if there are a whole row of these trees on a river-bank, the insects will all flash in exactly the same rhythm, and perfectly synchronized from one end of the row to the other – a quite fantastic spectacle in the tropical night.

Strangely enough, it is only males that are found on those trees, never females. They stay back in the jungle at night. So there is never a mass wedding here as in Jamaica. Then why all this split-second blinking along the riverfront? For whom are the beetles staging this splendid show? We don't know.

To heighten the effect and avoid mistakes in the nocturnal light of the marriage market, many tropical fire-flies, like deep-sea fishes, have their own specific pattern of light and colour.

The Mexican click-beetle, *Pyrophorus noctilucus*, is the insect shedding the brightest light in the world. On top of the thoracic shield it carries two little oval spotlights. Dr. Taschenberg[31] writes: "The use ladies make of it to add to

their attractions is very ingenious. In the evening they put
the beetles in a bag of fine tulle in the shape of a rosette;
and several of these bags are fastened to a dress. But as an
ornament the beetles are at their most beautiful when they
are worn on the hair in a wreath of artificial flowers made of
humming-bird feathers, together with a few diamonds."

Another Central American click-beetle wears two bright
white lamps in front on its "mudguards" and a red light at
its rear end; it is accordingly called "Ford bug." The illumination of the Brazilian beetle larva *Phrixothrix* is even
more sophisticated. It has two orange-red lights in front,
which glow like cigarettes. As soon as the insect gets frightened, it switches on a row of green luminous "windows,"
eleven each on the right and the left, so that it looks like a
train approaching in the darkness.

The illumination of German glow-worms is also a kind of
landing light for males intent on mating. In the dusk of warm
summer evenings the females of the species, *Phausis splendi-
dula*, which is fairly frequent in Germany, climb high grass
blades, displaying the luminous organs on the rear of their
abdomens. So that the light shall be visible from as far
away as possible for the males to be able to take their bearings, the females lift the back of their bodies so high over
their heads that the "luminous plaques" of their abdomens
point upwards.[32]

*Figure 21. A female glow-worm has its
hindquarters raised and emits fluorescent
light to attract a male.*

About half an hour later the males start a slow search-
flight at a height of between three and six feet. As soon as
one of them discovers the "lovelight" of a female on the
ground, he at once makes for it, hovering a moment or two
over his objective so as to improve his position, and then
simply drops like a stone.

To test their aim, Professor Friedrich Schaller,[33] Director of the Zoological Institute at the Technical College, Brunswick, carried out regular target practices with these luminous beetles. The objective was always a luminous female on the bottom of an open-topped glass jar, which was six inches high and had a diameter of only an inch and a quarter. The result surpassed all expectations: no less than sixty-five out of one hundred approaching males scored a direct hit. The others were never more than eight inches out in landing and immediately tried to reach the females on foot.

If a male has thus literally hit on a female, both extinguish their lights even before mating. The copulation is then performed, assisted by olfactory and tactile stimuli. But the strength of these instinct-releasers is less than that of light, for it often happens that while bride and groom are together and she has extinguished her light, he will suddenly see the light of another female nearby and quickly abandon his bride.

How strictly is the signalling code observed by these luminous beetles? To find this out, Schaller staged a number of interesting tests with dummies. With the other species common in Germany, *Lampyris noctiluca*, everything happened as one would expect: the strongest attraction for the male is an artificial light resembling the natural light of its female in all details, that is, brightness, colour, size, and shape of the luminous plaque. The more the substitute deviates from this ideal, the less the attraction. So there is hardly any danger that a male of this species will make advances to a female of a strange species.

Oddly enough, the *Phausis* males behave quite differently. They show a preference for blue light, which is not even present in the light spectrum of their females. Faced with a choice between a blue luminous dummy or a living female of their own species, they will plump for the dummy.

Furthermore, they are not even particularly fond of their females' luminous pattern. They prefer dummies with bigger and brighter lights and with an abnormally large number of luminous points. In short, these insects, "predisposed to infidelity," will make for any source of light, as long as it is neither too bright nor too large. Too much of these desirable qualities is counter-productive and causes flight reactions. Professor Schaller observed repeatedly how *Phausis* males landed with *Lampyris* females and then for hours made vain attempts at mating with them. Of course, nature does her best to make up for these shortcomings in another respect. *Phausis* has five times as many males as females, while with *Lampyris* the ration between the sexes is 1:1.

On the basis of this discovery, Professor Schaller ventured a prognosis: if in the future a mutational change in the hereditary factors of a *Phausis* female were to turn her yellow light into blue, the males of her species would immediately prefer her to all her other female rivals. The outcome would be that in a few years' time there would be only *Phausis* females with blue lights left. As the evolution theorists put it: the creature has a predisposition for such a development – a preplanned inclination towards future possibilities. In this case the basis for the predisposition is "merely" the way the insects' nervous system processes optical impressions.

Lighting engineers and power economists may well envy the luminous creatures their "light generators." The great mystery lies in the phenomenon of "cold light," which so far man has never been able to produce. While a man-made bulb converts into light only 3–4 per cent and a fluorescent tube 10 per cent of the energy supplied – that is, they act more like stoves than light producers – the little luminous beetles work at the ideal production rate of 100 per cent efficiency.

Only a short while ago the production of light entirely without friction or heat was considered Utopian, and the efficiency rate of luminous insects was therefore rated at about 90 per cent. But in 1961 it was proved that animal luminescence was, in fact, produced by perfectly cold light and that any quantity of energy supplied was converted into light. So biochemists are now trying to wrest from nature her technique of completely economical light production.

All we know so far is roughly this: in glands lying very close to each other the luminous beetles produce two chemical fuels – luciferin and luciferase. But this mixture does not shine by itself; other substances have to be added, first of all oxygen, as with every process of combustion. That is why a dense system of tracheae, pervading the luminous organs like respiratory tubes, provides for good ventilation. Magnesium ions have also to be present.

But the main thing is energy. It is supplied along the routes of innumerable tiny capillary vessels by, as it were, small tank trucks. The tank trucks are energy molecules swimming in the blood stream. Their chemical name is adenosine triphosphate, which is abbreviated into ATP.

In 1961, at The Johns Hopkins University, Baltimore, the American biochemist Professor William D. McElroy and Dr. Howard H. Seliger, [35], [36], [37] made crucial advances towards the understanding of these processes. They succeeded in working out the formula of luciferin's chemical structure and producing the substance synthetically.

The analysis of the second substance, luciferase, however, has proved much more difficult. This is a ferment or enzyme; that is, a protein macromolecule of complicated structure. Preliminary investigations have shown that it consists of about a thousand amino-acid units. So far biochemists have not been able to take apart many protein chains of that size, to record them, and put them together again, but they will no doubt manage it one day. We shall have to wait till then for our production of cold light.

But the basic process is clear: fortified by ATP-energy molecules, luciferase acts as a catalyst to remove two hydrogen atoms from the luciferin molecule and replace them by one oxygen atom (see Figure 22). Analytical chemists call this process dehydrogenation. As the hydrogen is passed on, light is radiated from the atoms' shells of electrons. At this juncture the American biochemists established an astonishing fact: the number of light quanta radiated out exactly equals that of oxydized luciferin molecules. This means that the efficiency rate is exactly 100 per cent.

At present our ideas about the glow-worm's "light switch" are still hypothetical. An ignition is obviously necessary to trigger off the process of light production described. The impulse for this is probably given by electrical nerve signals sent from the brain to the luminous organ. If a microelectrode is inserted into this nerve circuit and an artificial electrical stimulus provided, the beetles' lamp immediately flares up.

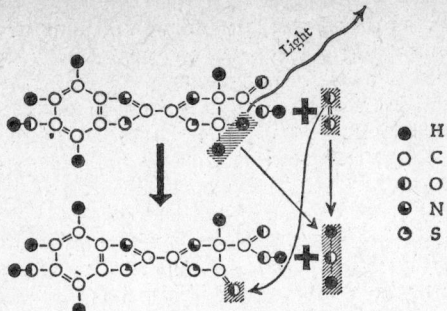

Figure 22. How glow-worms produce light. In the top half a molecule of luciferin can be seen, which consists of atoms of carbon (C), hydrogen (H), oxygen (O), nitrogen (N), and sulphur (S). If the molecule is acted on by luciferase as a catalyst and energy supplied by ATP (neither in figure) in the presence of oxygen, two hydrogen atoms split off, releasing energy in the form of a quantum of light being radiated. A dehydrogenated molecule of luciferin and water is produced (lower half of Figure).

The ignition is apparently provided by the nerve stimulant acetylcholine, which is always generated at the end of a nerve conductor just when an impulse arrives there. The details of the processes that follow are very involved and still need thorough investigation.

The Compound Eye

The eyes of insects and crabs, gleaming like gems with their many facets, strike us as strange and uncanny. Thousands of eyes go to make up one compound eye, every facet staring rigidly in its own direction, which differs by a few degrees from that of the facet next to it. For some decades scientists have been trying to find out what the world looks like to these creatures with a compound eye.

As long ago as 1900 Professor S. Exner[38] carried out a sensational experiment. From a luminous beetle's head he removed a tiny compound eye, substituted a film for the retina, and used this device as a miniature camera. This is one of the resulting pictures:

Figure 23. Is this what the world looks like to insects?

It shows a view of a church, seen through an arched window from the inside of a room. The letter R had been stuck to one of the window-panes.

Obviously, however, insects do not really see what this photograph seems to indicate. As with human beings, their optical image does not form on the retina, but in the brain; and, for insects as for the frog (see p. 29), we have to get away from our habitual idea that the eye's function is to convey a more or less sharp photographic image.

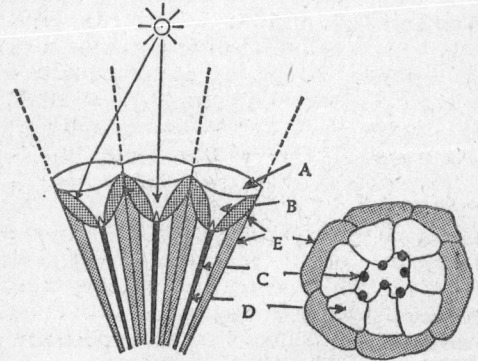

Figure 24. Three of the thousands of ommatidia of a compound eye. The light falls through the lens (A) and the crystal cone (B) on to the rhabdoms (C), which, like fibreglass optics, lead it down into the shaft and at the same time sideways on to the visual cells (D). Each ommatidium is screened off by light-proof pigment cells (E) from those next to it. Right: a cross-section of an ommatidium, which is shown to contain several visual cells.

A bee's eye, for instance, divides the sky into a screen of squares. Each of its 15,000 eye-facets, also called *ommatidia*, observes only its own sector of the picture, corresponding roughly to its opening angle of 2° to 3°[39, 40]. At each moment the sun is seen only by a single lens. This provides the insect with an almost perfect device for measuring the angle of its direction in relation to the sun, so that it can take course according to the sun's position.

In the same way the eyes of a bee in flight will divide the "map" underneath it by a screen. But the result of the process in the bee's brain is quite different: it creates a navigational instrument which until a few years ago was unknown to our aeronautics – a device for measuring its flying speed *over the ground*.[41]

Yet the principle is perfectly simple. There is no need for the insect to see every stone, blade of grass, or bush clearly; that would be much too complicated, and the perception of shapes is completely unimportant for it. It is enough for a single lens to register a change from light to darkness, and for the next or next-but-one lens to register the same change a short time afterwards. From the difference in time the bee's brain works out the flight speed over the ground. Our engineers have now copied this "invention by nature."

Man's perception of shapes, which to some extent is photographically correct, is an extraordinary achievement of the brain. It is a luxury a small insect could not possibly afford; far too much nerve substance is required. A bee has to, and does, solve problems important for it, with a brain smaller than a split pea.

It can determine its flying speed, for instance, not only over the ground but through the air as well. Its nervous system measures this through some sensory cells in the joint of its antennae: they register the bending back of the antennae by a head or side wind. Professor Herbert Heran[41] of Graz University, Austria, has discovered that nerve circuits in the brain compare the two speed values to work out the angle the bee has to fly at against the wind so as not to be carried off course, off the proverbial "beeline" from hive to feeding-place.

Another of the bee's feats of calculation at which the

human mind boggles is its determination of the angle at which to fly home after a long search-flight, which has led it in all directions. This is roughly how it is done. The insect takes an average of all flying angles in relation to the sun. In the process every single angle in its course is given a different value, which is according to how long the course was pursued. This produces the angle for a direct line from hive to the bee's present position. The angle has only to be turned by 180° to show the shortest way home.

I have simplified things, of course, to make the process easier to understand. For the bee doesn't really fly in a zigzag line but in curves. This increases the individual angles of its course to an infinitely large number, and a man would have to use integral calculus to work out the angle of its flight home. A mathematician has indeed already drawn up such a formula for bees' navigation with integral symbols.

No one need suppose that bees are experts in higher mathematics! After all, the human brain too performs mathematical feats it did not learn at school. My example above (p. 30) was that if light with a frequency of 385 million kilocycles hits the visual cells of the retina, we don't have to count them to know it is red light. A nerve circuit has already done the work for us, informing our consciousness merely of the result: red.

A similar thing happens in the bee's brain. There are circuits, like an electronic computer, which carry out the calculations automatically and at high speed, and eventually present the insect with the necessary plan of action. This results in the flying bee's "preferring" to see the sun only through one particular lens of its eye – in the direction it has unconsciously worked out.

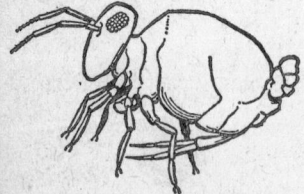

Figure 25. Spring-tail.

Is it the case, then, that insects can see nothing except a hazy change between light and dark, or at most the sun as a point of glaring light? This may apply with a few primitive insects, such as spring-tails and other microscopically small primitive insects that, instead of a compound eye, have only a raspberry-like assembly of a few eyespots or *ocelli*. More highly evolved insects have already taken a decisive step forward towards the seeing of shapes.

Ants, for instance, can recognize other ants at a distance of three quarters of an inch at maximum. Bees can tell blossoms from leaves slightly farther away, after being attracted by the scent to their immediate neighbourhood. Wasps memorize trees or bushes standing near the holes they have dug in the earth, so as to use them as landmarks when flying home. For the same reason one of the common sand crabs (*Ocypoda*) erects an approach-tower of silt, six inches high, close to the entrance to its burrow in the shallows.

Nevertheless, many experiments seem to show that these creatures do not see an ant, a tree, or a silt tower as we do, but only as an abstraction of significant characteristics and patterns, for which their brain has a specific receiving set – similar to the frog's described on page 32. The famous Munich bee expert, Professor Karl von Frisch,[42] found, for instance, that bees cannot distinguish between the following shapes:

Figure 26.

These shapes, too, apparently look all the same to a bee:

Figure 27.

Yet bees, strangely enough, do not have the slightest

Figure 28.

difficulty in telling apart these two shapes of flowers: although one would expect this to be a harder task than distinguishing between a circle and a triangle. The contradiction cannot be explained merely by their coarse-grained screen. In 1963 research by Dr. Rudolf Jander and Dr. Christiane Voss of Freiburg University's Zoological Institute[43] shed light on these questions for the first time. Here, for a start, are the facts:

A red ant, faced with a free choice of making for either a black dot or a shape segmented like a flower, will choose the dot: logically enough, since the ant has nothing to do with flowers, whereas the nearby entrance to its hill, which it has to head for, looks to it like a black dot. Just the opposite applies to a bee on a foraging flight: it prefers flower shapes to dark dots. But on its return flight, in a homing mood, it reverses its optical nerve system. Then it will look out only for the black dot, that is, the flight hole of its hive, and no flower can then lure it off course.

Similarly, if the caterpillar of the nun moth, *Lymanpria monacha*, falls off a tree trunk, it has to crawl up it again; so vertical dark beams, which represent tree trunks are the strongest attraction for it. The common hover-fly, *Eristalis tenax*, prefers vertical to horizontal lines, signifying the stems of plants it has to climb to find a good take-off spot. The water-beetle, *Stenus bipunctatus*, on the other hand, prefers horizontal lines to vertical ones, because its life may depend on being able to recognize the bank when in danger and to seek shelter there from attacking fishes. One might continue this list indefinitely, giving for each species of insects a specific spectrum of all the shapes for which it has a particularly developed receiving set. Everything else they either do not register at all, or else amorphously. As experiments have shown, this capacity for recognition is innate in the insects. Young, inexperienced insects react instinctively in exactly the same way as the other members of their species.

This inner necessity resulting from the activity of the nervous system, this completely unconscious compulsion to do a particular thing and that thing only, is in principle the secret of all purely instinctive behaviour in animals as well as man. There is something rather splendid about it; and perhaps intelligence, judgment, and understanding are no more remarkable than the "lifeway code" given to all creatures through the hereditary substance of their chromosomes, so that they can find their bearings on their journey through life.

Based on the findings above, Dr. Rudolf Jander[44] has put forward an important theory. He believes there may be nerve circuits in insects' eyes and brains that are similar to those that have already been shown to exist (see p. 22) in the nervous system of cats. Assuming for the moment there *are* these detector centres, they control the visual cells of the compound eye so as to help register and pass on information about dark or light dots, dark or light straight lines, straight edges, and certain directions of movement.

But in contrast to a cat's brain (and probably also to man's) insects have a much smaller number of such detector centres. In their small head there is room only for nerve chains taking in exclusively, say, vertical lines. Their brains simply have no provision for processing most of the lines running in other directions.

All this reminds us once more that man's capacity for absorbing a picture of his environment with many details is also remarkable. Indeed it is one of nature's greatest feats, achieved only through the use of hundreds of millions of nerve cells connected by complicated circuits which are still beyond our grasp.

However, [writes Dr. Jander], we can never stress enough that even human vision is full of faults compared to "photographic vision," as is proved by the innumerable optical illusions. We too, like the animals, see only a part of the world, even though a considerably larger part. But our vision is still far from being photographically perfect; for it to be so, we should need a brain perhaps three times the size, or even larger. Although we human beings may regard ourselves as the most perfect creatures on this earth, we should be modest enough to admit frankly that even we are still immeasurably far from the ideal of perfection. This applies to our visual abilities as much as to all our other attainments.

Eyespots

Besides its two compound eyes, the bee has three tiny eye-spots or *ocelli*. They are so small that you could never find them without a magnifying glass; they are in the hairy "fur" between the bee's two compound eyes. Their function was a complete mystery until recently; but in 1963 Dr. Burchard Schricker, a zoologist in Frankfurt,[45] and a pupil of Professor Martin Lindauer's, showed that they represent a kind of light-exposure meter.

The bee uses these eyes for the same purpose as a photo-grapher his light meter: to find out the amount of absolute daylight. Man manages very well without this faculty, except when taking photographs or using a moving picture camera, so why does the bee need exact data about the strength of the light?

It has to know when to fly out for the first time in the early morning, and for the last time in the evening. For the former, if it flies out too early, the light of dawn will not be bright enough for it to make out, even at very close quarters, the flowers it is seeking; if too late, it will lose valuable gathering time. Its absolute sense of light is even more important in the evening; for if it stays out too late, it will not be able on the flight home to recognize its hive, and will then have to spend the night in the open, exposed to mortal dangers.

This is why the bee's exposure meter reports the degree of twilight. If it is darker than ten times the light of a full moon, bees stay at home. Exact measurings, however, have yielded even finer distinctions: for instance, the last flight out in the evening takes place at a slightly brighter light than its start in the morning, for the bees reckon in advance the loss of daylight during their last flight of the day – an astonishing instinctive calculation of things to come, considering the smallness of a bee's brain. To put it another way, the light reserve is the bigger, the farther off the feeding-place is where the bee intends to gather honey. In a bee colony, therefore, the "long-distance transporters are likely to finish work earlier than the "short-haul" ones.

For a bee these eyespots are extras, as it were; but for other insects, such as spring-tails (see p. 55), caterpillars, spiders, and worms, they are the only light-receptor organs. The only thing these insects can "see" in their environment with their eyespots is a greater or lesser degree of brightness. An eyespot, however, also has certain directing characteristics, and several of them together may well work like a primitive compound eye.

The "Third Eye"

One of the most terrible figures of Greek mythology is Polyphemus the Cyclops with his single eye staring out from the centre of his forehead; such an eye also comes into

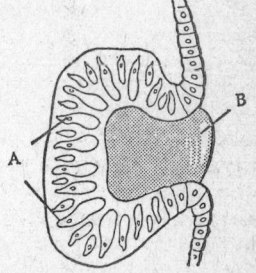

Figure 29. The ocellum came into being through an involution of the skin, in which there are cells sensitive to light (A). The gelatinous mass in the tiny cavity already represents a sort of primitive lens(B).

Grimms' fairytales. Were there ever men with this median eye? The question has been closely studied by Professor Eberhard Dodt[46] of the William G. Kerckhoff Heart Research Institute of the Max-Planck-Society, Bad Nauheim.

Certainly these legendary folk with median eyes are not the products of pure imagination. Doctors and zoologists know that every now and then children, as well as calves, dogs, and other animals, are born with a single eye in the middle of the forehead, rather large and as a rule grotesquely shaped. This malformation is known under the name of cyclopism, and it occurs when radiation or poisons have brought about a change in the genes.

But such a change causes other malformations in both human and animal embryos, and these have so far never allowed the new-born creature to live longer than a few hours. A day is the longest period known to science for a creature with a median eye to survive.

It is very remarkable, though, that the change of a single hereditary factor should produce such an odd phenomenon as a median eye. The phenomenon suggests that perhaps it was one of nature's construction models, which at some time in the earth's history somehow served a purpose, although with the forms of life prevailing today it has become useless and therefore obsolete. Indeed one has to go back a long way along the line of animal ancestors to hit on any real evidence for this; it is to be found with amphibians, fishes, and reptiles.

An impressive example is the New Zealand lizard *Sphaenodon punctatum*, called *tuatara* by the Maoris. This is in many respects a very strange creature, which by rights should have become extinct long ago. It lived on our earth 170 million years ago and is regarded as an ancestor of the dinosaur, or at least a close relative of his ancestors. But whereas the dinosaurs, those grotesque monsters with their enormous armoured bodies, have long disappeared from the earth, their ancestor still exists as a living fossil in the remote parts of New Zealand.

The *tuatara*, then, still has a regular cyclopian eye besides its two normal ones. In the younger specimens of these lizards one can still recognize it as a sort of glass tile, that is, a transparent horny scale. Immediately beneath it the skull too has a window. The strange organ even shows traces of a lens and a retina. Light-sensitive cells and ganglion cells have also been found here.

With advancing age the skin above the lizard's median eye thickens more and more. Eventually it grows so thick and dull that, doubtless, almost no light will penetrate.

But most true lizards also have a median eye. Its function is amazing: depending on the colour of the light which reaches the pineal gland through this eye, the gland produces more or fewer hormones; and these in turn cause a change in the reptile's skin so that after only a few minutes its colour resembles that of the light, producing an ideal camouflage.

In the blind-worm the median eye has a less unexpected purpose: because of it this lizard is not quite as blind as its name suggests.

The German zoologist Dr. Ingrid de la Motte[47] made some fruitful experiments on the "third eye" of trout and pike. She sealed up the fishes' ordinary eyes and put them in a big tank in complete darkness. Then she flashed a lamp and at the same time put food for the fish into a corner of the tank. After a few repetitions the fish would swim to the feeding corner as soon as the light was flashed.

This feat of training proves that the fish can register light even if deprived of their normal eyesight. As Dr. de la Motte says, they register light by means of their median eye. Even if the signalling lamp is dimmed more and more so that in the end a human being could hardly see its gleam, the fish still react promptly. This shows that their sense of light is a good deal more acute than someone might think who tried to imagine this perceptive faculty by standing in a room in bright daylight and "looking for" the window with closed eyes.

Seeing without Eyes

At the end of 1962 a strange report was making the rounds of the illustrated papers, daily press, and even scientific journals.[48] It was claimed that a young Russian woman, Rosa Kuleshova, could see with her fingers: that after members of a "testing commission" had blindfolded her, she could still read letters, and recognize drawings and colours, with the tips of the middle and ring finger of her right hand.

She and her testing commission have since been exposed as frauds: the miracle girl had been squinting past her eye bandage. Yet this supposed phenomenon was the subject of debate by reputable neurophysiologists all over the world for two years. They treated the claim seriously because science was familiar from the animal world with many examples of light sensitivity of the skin, a faculty of registering light without eyes.

Just as sensory cells sensitive to temperature are distributed over the whole surface of the human body, so fishes, worms, and clams have light-sensitive sensory cells in their skin. These are, so to speak, visual cells, like those in the retinas of our eyes, but without the ancillary structures, without lenses, iris diaphragm, or directional characteristics. They are simply and solely light-receptive organs which inform the brain about the degree of brightness.

Eels and lung-fish have a number of such light-recorders in the tail. This may seem a strange place for them, but it has, of course, its biological purpose. These fish like to hide in caves during the daytime, and they cannot see with their normal eyes whether the tail is protruding a bit from the cave, offering an attractive morsel for predators. The only way they can make sure about this is by the light sensitivity in the tail part of their skin. Even blind cave-fish have such a sensitivity to light.[50] This warns them against leaving their haunts of eternal darkness for the light-pervaded region of the cave entrance, where they might easily become the victims of sighted enemies.

With clams[51] this sense of light is so highly developed on the surface of the soft parts that the creatures can even register movements and recognize dangers, against which they can then protect themselves by quickly closing their shells or digging in. This is roughly how we have to imagine the process: just as we feel the touch of a finger passing over our skin, so the clam feels the light stimulus passing over the skin of its soft parts. Furthermore, the more numerous these light-sensitive cells, and the closer they are lying to each other in the skin, the finer the shades of light and darkness and of patterns the mollusc will be able to register.

If Rosa Kuleshova had had a skin like fish or clams, with an enormous number of light-sensitive cells, she would doubtless have been able to read letters with her fingertips. It seemed unlikely but not wholly out of the question that a human being could suddenly develop such a remarkable animal faculty; perhaps there had even been a mutation. This may have been what helped to keep the affair alive for so long, until the qualities involved were revealed as all too human.

Light sensitivity of the skin can anyhow be explained very easily by the presence of light-sensitive sensory cells in the skin. But there are also creatures without a trace of such cells which can nevertheless register light – a fact which science has long found amazing.

Considerable biological excitement has been aroused by the delicate little freshwater polyp *Hydra pirardi*,[52] which has a marked preference for light and will try as hard as it can to reach areas of its pond that are in the sun – although it has neither eyes nor light-sensitive cells. This happens in a quite extraordinary fashion: directly one of its arms is hit by a shadow, the polyp withdraws the arm abruptly. It does not make the slightest difference whether the shadow is cast by an enemy, which would like to have a nibble at the polyp, or by its own body, as is generally the case – the shadow being cast by an arm on the side facing the sun. The arms facing away from the light are then as if paralysed, so that the movement takes place only in one direction: towards the light.

In an equally baffling way, the beautiful sea anemone (*Actiniaria*), one of the cœlenterates which follow the arc of the sun with their crown of tentacles, can also register light. Dahlia anemones too (*Tealia*) must possess a mysterious sense of light, for they unfold their crowns only in darkness. Seapens (*Pennatula*), which in the darkness display their own ghostly green light, also notice when day dawns in the sea, and then switch off their illumination.

There are even unicellular creatures with a sense of light, for example, the *Amoeba proteus*, which lives in the mud of all stagnant waters. Directly it feels any shade, it moves along on its *pseudopodia* faster than before. If suddenly exposed to glaring light, granular streaming in the cell's endoplasm stops at once, and the amoeba stays motionless. But more, these creatures between half a millimetre and two millimetres in size can even distinguish two colours: red and a greenish blue. Their reaction is just the opposite to man's at traffic lights: they accelerate at red and slow down at green.

Things are more complex with some unicellular flagellates. They possess light-absorbing chlorophyll granules, so-called pigments. Sometimes these are even united as clusters shaped like pin-heads, so that they look like miniature eyes. Some flagellates carry little cell-wall domes over the pigment spots; that is, lenses that collect light rays. Strictly speaking, therefore, such unicellular organisms should not be called eyeless. There are also, however, many species of flagellates which, although they have beautiful eyespots, do not react at all to light and are apparently completely blind.

The flagellate *Euglena*[53] represents probably the acme of optical faculties among unicellular organisms. On the one side of its transparent body cell lies an eyespot of light-absorbing pigments, and on the other an opaque point, a sort of diaphragm. As soon as *Euglena* takes up a certain position in relation to the sun, the diaphragm shades the "eye." In this way the tiny creature can locate a source of light and make for the brightest place in its environment.

How can such tiny unicellulars possibly "see" at all? Dr. Rudolf Braun[54] of Mainz University's Zoological Institute suggests an interpretation based on chemical processes in the cells. We already know several processes that go on in a cell directly light falls on it: the degree of fluidity, the viscosity of the protoplasm and the permeability of the cell wall all increase; fats saponify; enzymes are inactivated (perhaps also activated); secretions of hormones are influenced, and many other things.

Light-absorbing pigments can probably intensify these chemical reactions and act as transformers of energy. For what purpose this energy is then used depends entirely on the structure of the chain of reactions in the cell's many "chemical factories"; and it differs from one species to the next.

But if many cellular processes can be influenced by light, this suggests that a certain "sense" of light is one of the basic faculties of the protoplasm and thus of life in general. The scale from *Euglena* right up to the human eye would then, in principle, represent merely a progressive refinement of this elementary faculty.

Colour Vision

It is generally believed that if you are chased across a field by an angry bull, the bull's anger is caused by a red dress or scarf or some other red object which irritates him. Even Spanish bull-fighters are firmly convinced that it is the proverbial red rag that drives the bull to fury. But animal psychologists are highly doubtful about this; for the acknowledged experts are still undecided as to whether a bull can recognize colour at all; more probably he sees the world only in black and white.

In this bulls are like many other animals,[55] including some lemurs, raccoons, golden hamsters, field mice, and marsupial rats. Scientists also disagree on whether dogs, cats, rabbits, rats, and housemice are colour-blind or not. In any case their ability to distinguish colours is so small that it can scarcely be proved.

When an alligator[56] is given a choice between two cardboard disks and fed only if it chooses the coloured one instead of the grey one of equal brightness, it begins to feel extremely insecure. In its mental conflict, after a few repetitions it will even develop an experimental neurosis, showing indifference not only to cardboard disks, with colours it evidently finds hard to register or else meaningless, but to food and rewards as well. It will crawl into a dark corner and hide there for a few days without moving a muscle.

For hedgehogs there is apparently only one colour in the world, and that is yellow. The red-backed vole has a sense only for yellow and red, the small civet cat only for red and green. All other colours of the rainbow are just as non-existent to these creatures as ultra-violet rays are to us. Yet the few colours such creatures see appear to play an important part in their lives.

We know of the frog[57] that if in danger it will make indiscriminately for something blue, anything from the water of a pond to a sheet of paper proffered by the experimenter. At moments of apprehensive excitement the frog feels repelled by green, though clearly without awareness that a jump into greenery – the grass, for instance – would increase the danger of its being eaten. But nature has synchronized the frog's senses and instincts, as a substitute for reason, so that it automatically does the right thing.

A similar thing happens with some species of aphid,[58] or green-fly. If the weather is fine, these winged vermin will embark, literally, on a journey into the blue. They simply rise upwards and allow themselves to drift in the wind for a few hours. Then their eye switches from blue to yellow – that is, onto the yellow colour component of young shoots. They thus lessen the danger of a false landing. Rose-growers exploit this colour orientation in the reverse sense: they paint the bottoms of some bowls of water with a special yellow and put them on their rosebushes as aphid traps.

Horses, red deer, sheep, giraffes, squirrels, guinea pigs, and polecats can grasp a wider section of the colour spectrum. Dr. Gerti Dücker[54] of Munster University's Zoological Institute writes:

> The sense of colour is probably best developed in monkeys. A full appreciation of colour, as with man, has been proved for *Catarrhini* like the rhesus monkey, pig-tailed macaque, baboon, Java monkey, and long-tailed monkey, but also for the common marmoset, which belongs to the *Platyrrhini*. With the other *Platyrrhini* investigated, such as the capuchin monkey, spider monkey, and squirrel monkey, a shortening of the visible spectrum at the red end became apparent, so that their colour vision could be compared to that of a red-blind human being. In the chimpanzee sensitivity to colour differences is at least as good as in man.

Man can distinguish 250 pure colours, from red by way of orange, yellow, green, blue, and indigo up to violet, and about 17,000 mixed colours; plus about 300 shades of grey between white and black. (A bee's vision comprises only twelve such shades and that of the fruit-fly *Drosophila* a mere three!) So our optical sense can take in five million shades of colour altogether – truly one of nature's master achievements.

What emerges from a comparison with the animals is that man's colour vision, like the way he sees an image, is no matter-of-course affair, for it requires a sensory and nervous system that is little short of miraculous.

How Does the Eye See Colours?

Astonishingly enough, up till 1962 no scientist could tell the mechanism by which the eye and brain recognize colours. There were many theories, all postulating the existence of at least three different colour-sensitive cells; that is, cones. But to make nonsense of all the theories, despite an intensive search neurophysiologists could trace only one type of colour receiver and only one kind of visual purple. Today we know that these negative results of their investigations were solely due to the inadequacy of their measuring instruments.

For in 1963 something happened not uncommon in the history of science: the same discovery was "in the air" and was made almost simultaneously by several scientists. In this case four groups of research workers arrived independently at the same result: Professor Hansjochem Autrum,[59] head of Munich University's Zoological Institute, Professor George Wald of Harvard University and his group, including Mrs. Wald and Dr. Paul K. Brown; Professor Edward F. MacNichol, Dr. W. B. Marks, and Dr. W. H. Dobelle of The Johns Hopkins University; and a British physiologist, Professor W. S. Stiles, then of the National Physical Laboratory in Teddington.

The result: in the eye of full colour-comprehending creatures there are, after all, three different types of colour-vision cell. Each type has a different visual pigment, and so each type of cone responds to a different section of the colour spectrum. Edward F. MacNichol,[60] Professor of Biophysics at The Johns Hopkins University, Baltimore, found for man the following maxima of sensitivity in the three colour receivers:

Cone type A: deep blue-violet – 450 nanometers wavelength
Cone type B: dark green – 525 nanometers wave-length
Cone type C: dark yellow – 555 nanometers wavelength

Nature's selection of these basic colours is surprising, for it contradicts previous assumptions that all shades of colour would be mixed in the eye from the basic colours of red, green, and blue, just as in colour television. Yet from the colours nature has chosen for the human eye, all other colours can be put together, provided the type of visual cell is also sensitive to red. As emerges from figure 30, the cones with maximum sensitivity in the yellow section also respond to bluish-green, green, and red light. If they alone are stimulated, and no other cones at the same time, it is the unequivocal signal to the brain for the sense of "red."

This certainly does not mean that in all colour-sensitive

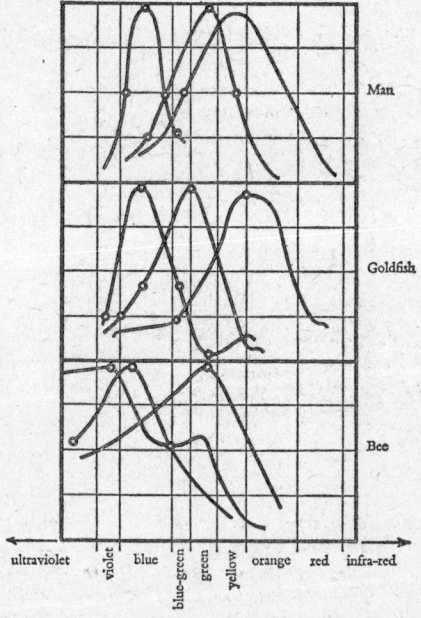

Figure 30. The different ways that man, the goldfish, and the bee see the colour spectrum. Each graph shows the sensitivity to light of one of the three types of cone with light of different colours.

animals the colour sections in the types of visual cells are distributed over the spectrum exactly as with man. Visual cells are not nature's uniform and universal building blocks. For instance, in the rhesus monkey and the goldfish the characteristics of sensitivity are slightly displaced, and in the bee a good deal displaced.

A bee's vision has the basic colours green, blue, and ultra-violet, the last of which is invisible to us; but red is no colour for the bee, merely an ineffective section of the spectrum, which it sees as black.

New research is already shedding some light on the causes of colour-blindness. One might assume, for instance, that a man unable to see some colours lacks one of the three types of colour receiver, or the visual pigment belonging to these cells. This is quite possible, but not certain, for there might

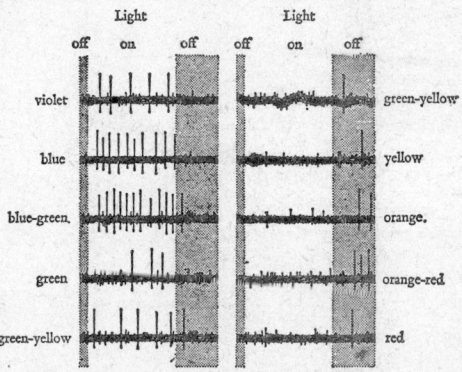

Figure 31. The secret of complementary colours is expressed in this Figure. Complementary colours produce in the brain the impression of (grey) white, when they blur into each other by being rotated quickly on a wheel; while the mixing of the same colours on a painter's palette produces other results. This Figure shows the "signalling activity" of a ganglion cell, when the appropriate visual cells are exposed to light of different colours. The strongest excitation occurs with blue-green, the strongest inhibition with orange-red. These are two complementary colours. When they light the retina in quick alternation, excitation and inhibition cancel each other out – which produces the impression "white".

also be a defect in the processing of the stimulation signals in the higher nervous system.

A cone lying in the retina of the eye, of course, has no "private" nervous line to the brain (see p. 19); the sensory impressions are sorted out and collected on the way according to different viewpoints by the bipolar cells and the ganglion cells.

Several neurophysiologists[61] have already gained some insights into this "magic garden of the inner rainbow." As with excitation fields described earlier, they discovered ganglion cells, which are excited by a group of yellow-sensitive visual cells and inhibited by a group of green-sensitive visual cells. Other ganglion cells react the other way round: yellow light in their excitation field slows down the volleys of electric impulses that they fire towards the brain, while green light accelerates them. Other cells pass on only stimuli from a definite narrow section of the spectrum, and yet others seem entirely independent of colour, register-ing only brightness or darkness (see p. 23).

According to present knowledge, then, colour vision is a process of at least two stages. The visual cells work on the three-colour principle. But the succeeding nerve chains codify the information received partly on the principle of complementary colours: these colour nerve-conductors to the brain are polarized on two colours, one with excitatory, the other with inhibitory effect. The resulting signal code for the various colour sensations is shown in figure 31.

The "one-colour messengers" also take part in the build-up of the total colour impression in the brain. In these nerve circuits, of course, defects may occur as well. Depend-ing on the source of error, a particular type of colour-blindness will result.

This leaves some important questions unanswered. How does light stimulate the visual cells? How does the brain form a coloured picture of the information coming in? How is the impression of mixed colours produced? At present we do not know.

Seeing Ultra-Violet Light

"Anyone who thinks the whole floral splendour of the earth was created to delight the eye of man should study the colour sense of the bees and the quality of the flowers' blossoms, and he will learn modesty." Such is the admonition of Professor Karl von Frisch,[62] who has so impressively made accessible to us the totally different colour world of the bee: a world with no red; with the stars of daisies, which to us appear white, looking bluish-green; with white roses, apple blossoms, morning glories, and bluebells shining in all sorts of different colours.

> If the petals here owe their colourful garb to the absence of ultra-violet light [Frisch goes on], in other cases the addition of it is the source of a colour magic hidden from us. The yellow blooms of treacle-mustard, rape, and charlock, for instance, are for us hardly distinguishable in colour and shape; but the bees know better. To them only the treacle-mustard is yellow. The rape blossoms reflect a little ultra-violet light as well, which gives them a slightly purple tint. The petals of charlock reflect a lot of ultra-violet, so charlock looks deep crimson to the eyes of a bee, which demonstrably can easily distinguish between all three species.
>
> Anyone who could look at the world through a bee's eyes would be surprised to discover more than twice as many kinds of bloom as our ultra-violet-blind eye can see, with ornaments never registered before.

Why does nature lavish so much red on the decoration of blooms, when bees cannot recognize red? There are several reasons. To us a poppy radiates pure red; but if we examine it more closely, we find it reflects ultra-violet rays as well. So to the bee it has a pure ultra-violet colour. The same sort of thing applies with heather, red clover, and the Alpine rose. Their red is not pure, but a crimson mixed with blue; so to the bee their blooms appear blue.

In the tropics, however, there are flowers of a pure red colour which do not reflect any ultra-violet; they confirm the general evolutionary idea of colour vision. For bees simply ignore all these flowers: this red is not "meant" for bees. But it applies to the "red-sighted" humming-bird, which hovers in the air like a helicopter and sucks the nectar from the flower's calices.

In Europe too there are red flowers without any admixture of ultra-violet light: sweet-william, campion, and catchfly. Their bright red colour is also not meant for bees but for daytime butterflies, which have a very good receptor for red. This seems highly logical. One might almost work out from the colour combination of the bloom whether its visitors are sensitive to red or not.

A special kind of deception is played on us by the Indian luna moth (*Actias luna*).[63] To our eye both male and female of this butterfly are light green and indistinguishable from each other, and to their enemies they are hardly visible against the green background of the leaves. But to their own ultra-violet-sighted eye the picture looks very different. To them "she" looks fair and "he" looks dark. Nor is the green a camouflage colour for them; against the greyish green of the leaves they see each other as brilliant spots of colour.

So there are eyes which can see the colours of the spectrum visible to us, and also ultra-violet. But besides that there is the horseshoe crab (*Limulus polyphemus*)[64] living in American coastal waters. As its Latin name indicates, it has a "third eye" on top of its "steel helmet" between its two compound eyes.

The third eye, on closer observation, consists of two eyes adjoining each other. The odd thing is, that these are not also compound eyes, as one would expect in an arthropod, nor are they ocelli, as with a bee, but camera eyes – very tiny ones, it is true, half a millimetre in diameter each and with only fifty to eighty visual cells in the retina; but still camera eyes.

The horseshoe crab, about a foot long, with its hard armour, is a living fossil, which populated the earth 160 million years ago in the Jurassic period, and is probably a near relative of the common ancestor of all crabs and spiders. Zoologists had long wondered why the creature should need this "invention before its time" of two camera eyes. In 1964 Professor George Wald of Harvard University established that the horseshoe-crab's third eye is a receptor for ultra-violet light. So with one pair of eyes it sees the

Figure 32. The solidly armoured horseshoe crab, a foot long. The view from above (top) and the front view (bottom) show the position of the compound eyes (A) and the camera eyes (B).

ultra-violet section of the colour spectrum and with the second pair of eyes a different section of that spectrum – an amazing division of labour! We still do not know, however, why it should need a special receptor for ultra-violet. The reason may be to enable it, like the bee, to make out the sun even through clouds.

For bees,[65] even though the whole sky is overcast with a bank of clouds, can take their bearings from the sun's position. How is this possible? Before the Second World War Karl von Frisch had already made experiments with light filters and concluded that bees could recognize the sun's ultra-violet rays through the clouds. But the physicists refused to accept this, as they had established through photographic tests that a bank of clouds completely absorbs ultra-violet light!

So the bees' ability to take their bearings from an overcast sky remained a mystery for a long time, until in 1959 new, highly sensitive photographic plates began to be marketed. Again a camera was pointed at the place where the sun must be hiding behind clouds. Pictures were taken, and now man too could for the first time photograph an ultra-violet illumination of the sun's position. The fact which had long been intimated to biologists by the bees belatedly found a brilliant material confirmation. But if the cloud formation is so thick that even ultra-violet rays cannot penetrate it, the bees, too, will be at a loss and forced to take time off.

Chapter Two

The "Temperature Sense"

The Rattlesnake's "Third Eye"

In 1952 the American neurophysiologist Professor T. H. Bullock staged an exciting experiment[1] at the University of California. He put a rattlesnake and a mouse together in a small terrarium, sealed up both the reptile's eyes with adhesive tape and sprayed the inside of its mouth with a chemical fluid which blocked the olfactory nerves. Would the snake be able to find its victim even without these senses?

Bullock waited tensely to see what would happen. After only a few minutes it became clear that the rattlesnake had no intention of playing blind-man's-buff with the mouse. It seemed to know all the time precisely where the mouse was running. Intent on its purpose, the snake followed the mouse, coiled itself up like a steel spring, jerked forward, and at once secured a direct hit. Ten minutes later half the mouse had already disappeared into the snake's distended mouth.

Although the reptile had been unable to see its prey, smell it, hear it, or feel it, it had located the mouse with the assurance of a sleep-walker. Was this due to unknown rays or a mysterious sense inaccessible to human beings?

When Bullock made a closer study of the snake's body, he discovered two small dimple-like pits, placed like a car's headlights at each side of its head between nostril and eye.

Perhaps this solved the mystery. Now he sealed off these pits too with adhesive tape, again put the snake in the terrarium, extinguished the light, and put at least a dozen mice

Figure 33. The head of a rattlesnake seen from in front. Between the eye (B) and the nostril (D) lies the heat-ray eye (C). (A) represents the tongue.

in with it. Even after a few days the rattlesnake had not caught a single mouse! Only closing the two dimples had really "blinded" it.

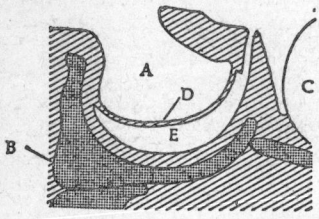

Figure 34. The rattlesnake's heat-ray eye lies in a deep pit (A) with an entrance about three millimetres in diameter. It is embedded between lower jawbone (B) and the regular eye (C). In the reflector-shaped membrane (D), there are 150,000 nerve cells sensitive to heat. The interior (E) is connected with the outside world by a narrow duct.

What strange sensory organ may be hidden in these dimples? In the thick skin lining these pits, Bullock found sensory nerve cells which were in fact less strange than he had expected. For they also exist in human skin in a very similar fashion and enable us to feel warmth. But while man has only three of these heat points per square centimetre of skin, there are no less than 150,000 heat-sensitive sensory nerve cells crowded together on the same area of the rattlesnake's pit organ – five times more than the total that man possesses.

This concentration of heat-sensitive cells gives the rattlesnake's sense of temperature a gigantic boost. By this means it can "see" heat rays, and that applies not only to a very hot stove, but also to the rays of the body temperature of a mouse or of other creatures or objects, as long as they differ from their environment by even a few tenths of a degree Centigrade. If we men had such infra-red eyes, a mass meeting at night would look to us like a swarm of glow-worms of outsize proportions.

Moreover, the pit organs are formed like reflector mirrors, so the rattlesnake can receive heat rays only from two sharply defined cone-shaped regions. If it moves its head this way and that, besides registering the presence of something warm, it can feel out that something's approximate size and shape. Probably this is how it tells a rat from a snake-eating mongoose.

But if in the darkness the rattlesnake is offered, instead of a mouse, a light-bulb switched off but still warm, which is wrapped in material, the snake will pounce on this as well. The heat-ray image it receives of its environment must therefore resemble a view through fogged-up glasses.

Incidentally, the "third eye" is also a great advantage to the snake when hunting in the daytime. A number of lizards and salamanders can adapt their shape and colour deceptively well to their environment, so the snake cannot make them out with its ordinary eyes, but the heat ray discovers them at once.

This example shows that senses which at first seem mysterious, "metaphysical," may be based on perfectly normal and universally known physiological facts. The extraordinary eye which can see heat rays is merely an assembly and functional arrangement of things long familiar.

Heat-Sensitive Organs in Insects

Two writers have very recently advanced the view that rattlesnakes and other pit vipers are not the only creatures to have heat-ray eyes. Lorus Milne[2] believes that mosquitoes, besides being attracted by the smell of humans and animals, can take their bearings from the heat rays emanating from their victims. And Dr. Philip S. Callahan[3] reports on infra-red-sensitive organs in night moths: through this "ultra sense" the males can apparently find their way to the females on dark moonless nights, "if necessary, against the wind and the smell."

Professor Ferdinand Porsche, a car designer noted for practical sense, once said to his students, "The best thermometer is still the hand," and directly afterwards burned himself on an exhaust pipe. This is a striking illustration of the fact that man's sensory organs are far from precision instruments; such an accident, for instance, would never have happened to a bed-bug!

Some creatures, indeed, have a sensitivity to heat and cold vastly superior to ours. When blindfolded – playing blindman's buff, say – we cannot pick out people yards away by the heat difference in the air. But for the bed-bug this is one of life's necessities, as it climbs up to the bedroom ceiling at night, to feel with its antennae for the heat source of the sleeper. After following every turn its victim makes, it finally drops on some naked part of his body with the precision of a bomb.

A body-louse too, if its eyes are sealed off to shut out any light, will keep steadily running, at a few inches' distance, after a human finger or other warm object. Professor Konrad Herter,[4] the Berlin neurophysiologist, has worked out the minimum temperature difference that a mosquito can still register: one five hundredth of a degree C. at a distance of two fifths of an inch in the air.

Heat Sensitivity in Fishes

Fish too are at least equally sensitive to heat. If their heads touch water only 0·054° Fahrenheit (0·03° Centigrade) warmer than the water the rest of their bodies are immersed in, the fish notice it immediately. Compared to them, we human beings are more like insensitive pachyderms. The capacity of noting minimal fluctuations of temperature is vitally necessary for fish; it saves them from straying into cold ocean currents or deep waters.

Moreover, their extraordinarily well-developed temperature sense probably makes them capable of regular underwater navigation. It is part of the sole's life cycle, for instance, that it is newly hatched in the open sea, moves into shallow coastal waters at the beginning of April, then transfers again to deeper waters over softer sandy ground. How do these flatfish at the bottom of the sea know which direction leads them to the shore and which to the open sea?

A German zoologist, Dr. Manfred Zahn,[5] after many years' experiments at the Heligoland Biological Station, thinks it most probable that the sole in their migrations take bearings by their sense of temperature. In experiments with aquaria warmed to different temperatures, he found out the following: at the end of March and beginning of April the fish prefer to stay in water with a temperature of about 68° F. (20°C.). This temperature they can find only in coastal waters which are quickly warmed by the spring sun. So they have simply to swim with a current of colder water to be "sure" this is the tide which will carry them to warmer regions. Directly they feel the warmer ebb-tide current, they dig themselves into the sand and wait for the next tide.

At the beginning of June they have a peculiar "change of mind." Suddenly they no longer like the warm water, and start showing a preference for a colder water temperature of about 57·2° F. (14° C.). They can still distinguish warm ebb-tide currents from the cold tide; only now they do the opposite, letting themselves be carried by the ebb-tide back into greater depths. So far, however, we know nothing about the indications by which they recognize the seasons so exactly and adjust their time-tables.

Absolute Sense of Temperature

Many animals have another faculty far more highly developed than in man: the absolute sense of temperature. Let us imagine a man walking through a museum, say, with many rooms, all at different temperatures, that is, 61°, 63°, 65°, 67°, 69°, and 71° F., but not in any recognizable sequence. If the man going through them had to tell by his sense impressions which was the room with the 67° F. temperature, he could only guess; whereas many animals, having an absolute sense of temperature, would know which it was without any difficulty. Just as a musician with perfect pitch immediately recognizes a note, as for example, C sharp, so rodents, bees,[6] and fish,[7] if previously trained to do so, will recognize a temperature of 67° F. with complete accuracy to within a degree, irrespective of whether they enter the test room from a warm or a cold environment.

The Incubator Bird

A peak of precision is found in the Australian brushturkey (*Leipoa ocellata*), also called incubator bird, one of the megapodes or mound-builders. It builds itself an incubator, scratching together some leaves and grasses, and hatches the eggs in the warmth of their decomposition.

In the egg chamber of its mound, which is two or three feet high, a temperature of 91·4° F. (33° C.) must always be maintained. This involves a tremendous strain for the bird every day for six months, beyond anything a human being could be expected to bear. Depending on whether it is hot or only warm in the Australian bush, whether it is day or night, whether the sun is shining or hiding behind clouds, whether the fuel collected is supplying plenty of heat or is nearly used up – depending on all these things, air vents have to be dug or closed, a covering of sand as heat insulation has to be removed or put on, thickened or thinned out; and sand which has been cooled by the shade or warmed by the sun must be either taken into the air-conditioning chamber near the eggs or removed from it.

Dr. H. J. Frith[8], an Australian zoologist, made life even harder for the bird: he built into the mound three electric heaters supplied by a Diesel generator 100 yards away, and switched them on and off at random. The incubator bird became terribly agitated, as if unable to understand the way the temperature kept rising and falling. Yet by purposeful action it did exactly the right thing to keep the egg chamber at 91·4° F. (33° C.). Dr. Frith never managed with his heaters to change the temperature in the mound faster than the bird could correct it.

Every few minutes the bird would stick its beak into the

Figure 35. View and cross section of the heated mound of the incubator bird. Below the egg chamber is the rotting compost, above it the layer of sand three foot thick as heat insulator.

mound in different places and withdraw it again, filled with sand. It then let the soil sample slowly trickle out at both sides, after testing the temperature with the "thermometer" in its tongue or the roof of its mouth. Its sensitivity is so highly developed that it would gauge the temperature inside the mound to within a tenth of a degree and then act on its findings, with as much precision as if it had studied thermodynamics.

Excitation of Nerves by Temperatures

An experiment with an electric friction machine, favourite toy of physics students, gives a first insight into the strange world of the temperature senses. If you let the sparks fly to different places on the skin, you will sometimes feel warmth, sometimes cold, sometimes pain, sometimes a sense of touching something, depending on which type of sensory nerve has been hit by the spark.

This shows an essential point: that there are two different temperature receptors in our skin – heat recorders and cold recorders. Neurophysiologists have studied them more closely by tapping individual nerve fibres, in the same way that I described for listening in to the transmissions of the visual nerves (see p. 22). What signals do the temperature-sensitive organs send to the brain?

As Professor H. Hensel[9] found with the cat, cold receptors at constant skin temperature transmit code signals with fair regularity. As soon as the skin gets warmer, they spontaneously stop firing. But if the temperature drops, they send their code signals faster the more it cools down. Heat receptors work just the other way round.

And here is a strange thing: as listeners-in to the transmission of the nerves, we can immediately recognize something our brain can't: the absolute temperature of skin. At 77° F. (25° C.) a heat receptor always fires once a second and a cold receptor ten times a second. If the cold receptor has stopped and the heat receptor sends three and a half signals per second, you can safely say that the skin has taken on a temperature of 96·8° F. (36° C.).

So our brains are always being exactly informed about the skin's absolute temperature. Why, even so, do we have no absolute feeling for temperature? Is it that our brains lack an inner stop-watch to measure the speed of the arriving signals? Or is it that the signals on their way to the brain are processed together with other nerve signals? We do not

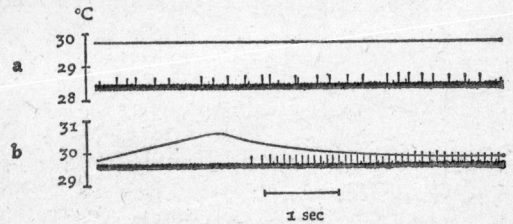

Figure 36. How a cold-sensitive cell works. In constant temperature (top thin line in graph a*) it sends 'interval signals' in slow succession (bottom thick line with impulse notches). If the temperature rises (rising part of the thin line in graph* b*), it stops the transmission. But it reacts to dropping temperatures with fast volleys of impulses.*

know the answer so far, but at least we are learning how to explain the absolute temperature sense of many animals.

Three more oddities are noticeable if a graph is drawn up, as was done by Dr. Y. Zotterman,[10] for the signalling speeds of heat and cold receptors at all temperatures between 50° and 132° F. (10° and 50° C.).

Remembering the sensitivity curves of the colour-sensory cells, one might say, "Just the same idea." If possible, indeed, nature always works on similar principles. There the firing speed of the nerve cells depended on the wavelength of the light, and here it depends on the temperature. With colour sensation there were three different types of sensory cells; here there are two.

Moreover, the cold receptors, as can be seen from figure 37, stop their transmission altogether above a skin tem-

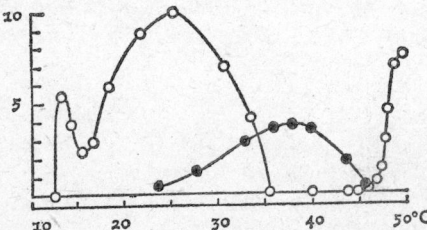

Figure 37. Volleys of nerve impulses showing reactions of cold fibres (circles) and warm fibres (filled circles) to different temperatures. The vertical scale shows how many impulses are sent per second.

perature of 95° F. (35° C.), leaving the field entirely to the heat receptors. The temperature of the surrounding air or another medium may, of course, be much higher than the skin temperature, which has intermediate values in relationship to the environmental and body temperatures. Above 113° F. (45° C.) these in their turn fall silent, but strangely enough, the cold receptors start transmitting again. That is why with severe burns we feel cold instead of hot a "paradoxical cold sensation", as it is called. We do not yet know, however, why the nerve cells behave so oddly. At present, indeed, we know next to nothing about the mechanism and biochemistry which change heat – that is, the kinetic energy of molecules – into electric nerve signals.

With every species of animals the temperature curves of the heat and cold receptors – like the characteristics of the colour-sensitive cells – vary to a greater or less degree. While the human cold receptor can at most send sixty impulses per second to the brain, the cold receptors of animals with especially good temperature sensitivity are capable of firing at a good deal higher speed. In a ray (the fish), for instance, they fire up to fifty times a second if the skin temperature

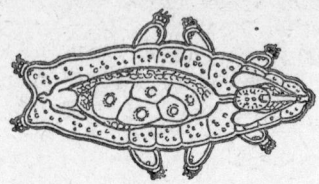

Figure 38. The bear animalcule.

remains constant for some time. But if it cools down rapidly the signalling speed will even soar to 200 impulses a second. Temperature differences of as little as the twentieth part of a degree already have a distinct influence on the signal pattern. This is the whole secret of why the sense of temperature in animals is so very highly developed.

Favourite Temperatures

There is a minute animal called a bear animalcule, which looks like a surrealist caricature of a fat eight-legged bear. It has a reputation as a "clown in the microscope circus," and lives in moss, where it wraps itself up into the shape of a barrel and can survive extreme climatic fluctuations quite unscathed. Some of these creatures were subjected to a "torture chamber" experiment to determine how extreme the temperatures may be. Ralph Buchsbaum and Lorus Milne[11] describe it thus:

> For an hour bear animalcules were exposed to a stream of hot air of 197·6°F. (92° C.). When they were moistened afterwards at room temperature, they were lively again half an hour later! They were fused into glasses lacking any oxygen; they were kept for months in pure

hydrogen gas, pure nitrogen, pure helium; in carbonic acid, hydrogen sulphide, coal gas – it did not do them any harm. The shrunken mummies woke up again every time as long as they received water, which is their livelihood. For twenty months the little barrels lay in liquid air at about minus 392° F. (minus 200° C.); afterwards for 8½ hours in liquid helium at an outer space temperature of minus 519·8° F. (minus 271° C.) Yet they thawed out again and became as lively as before.

This is the utmost any creature on earth can stand. There are, however, a number of other creatures which not only survive in temperatures that would be lethal to man; they even thrive on them. These creatures range from the fish of the Hebron Fiord, North Labrador, which have to exist all through the year in supercooled water of 30·4° F. (minus 1·7° C.), to the snow-flea, and the scorpions of the desert, which come fully alive only at a temperature of 113° F. (45° C.).

Man feels best when his skin has a temperature of 91·4 F. (33° C.), which it will normally be when the air temperature is between 60·8° F. (16° C.) and 71·6° F. (22° C.). These temperatures strike us as neither cold nor hot; they are a "comfortable warmth." It is odd that we say, "It's nice and warm," when in fact we do not feel any warmth at all. So, basically, the registering of cold or warm is a signal to warn us that we ought to do something to restore the state of comfort: throw a log on the fire or put on a pullover. But an animal has no other choice than to look for a place nearer its preferred temperature. In the graph of figure 37, this lies at the intersection of the curves of heat and cold.

Animals stay where the temperature is most invigorating for them. The favourite temperature for each species can be found out in a temperature-choice apparatus, a long narrow space with a temperature gradient from one end to the other. Whether the subject is a snake, a goldfish, or a parrot, they will at once pick out the place in the apparatus where they feel comfortable.

Preferences differ a lot:[12] the snow flies living by the

glaciers of the Tyrol like 39·2° F. (4° C.) best; winter moths prefer 39·2° F. (6° C.), rainbow trout, which live in cool mountain streams, 50·72° F. (10·4° C.), carp 70·34° F. (21·3° C.). Fresh water crabs of the Mediterranean countries, which enjoy lying in the sun, like 86° F. (30° C.) and ant lions in the hot sand of the Libyan desert 120·38° F. (49·1° C.).

African witch doctors use knowledge of an insect's favourite temperature rather like a clinical thermometer: the head-lice in their patients' hair are extraordinarily sensitive to temperature, and as soon as the patient's temperature gets too high or too low, the lice leave him for healthy people. So people without lice must be ill. Unfortunately, lice and fleas leave not only sick people but sick animals, including disease-carrying rats. That is how fleas spread plague and lice spread typhus with the appalling speed that used to terrify the world in past centuries.

We have already seen from the sole that a creature's favourite temperature may alter with the change of the seasons. The same sort of thing happens with the bark beetle. In spring and summer it is attracted by the warm sun, which, preferably, should be between 80·6° F. (27° C.) and 91·4° F. (33° C.). In the autumn the insect changes its mind, suddenly feeling comfortable at only 44·6° F. (7° C.). This makes it look in good time for a secluded spot for hibernation before the severe cold sets in. The tiger beetle, indeed, changes its taste in temperature twice a day: in the morning its longing for warmth drive it out of its sleeping hole, and in the evening its desire for coolness makes it "turn in."

The dreaded tsetse fly's behaviour is even more decisively influenced by temperature sensitivity. As long as the temperature is bearable, it instinctively seeks brightness; that is, it flies out of the bush onto the grasslands where its victims are grazing. But directly the thermometer rises above 86° F. (30° C.), when cows and antelope are looking for shade, the tsetse fly, which is shortsighted, switches over, now heading for anything that looks dark. It is thus bound to find its victims again.

On the basis of this knowledge Professor Konrad Herter constructed a tsetse-fly death trap. He lit part of the test room brightly, heating it to 107° F. (42° C.). As soon as the insects were let into the bright part, the temperature made them want to fly into darkness (negative phototaxis). They unhesitatingly obeyed their instinct, although they were slowly killed by the heat in the dark part of the room.

The inner compulsion, although it had lost its original purpose, was in this case stronger than any urge to escape the scorching heat. So strange are the ways in which animal behaviour may be controlled through sensory impression processed by the structure of the instincts.

The Thermostat in Man

No electronically controlled air-conditioning system in the world keeps a building so exactly at equal temperature as "natural" central heating keeps the human body. Without this regulation we could never maintain in ourselves a constant heat of 98·6° F. (37° C.) in summer and winter, at the North Pole as well as in the sun-drenched Sahara. Without it a cool evening or rainy weather would make us stiff and cold like a frog, while ice and snow would make us rigid like a piece of frozen meat. We should have to go into hibernation for several months.

The "invention" of a central heating system that is adjustable over a wide range is one of the most magnificent achievements in the evolution of the vertebrates. In contrast to the cold-blooded animals, with their body temperatures at the mercy of the temperatures all around them, man's central heating keeps the life forces always at optimal functioning heat. How does it manage this?

For any central heating system, in a house or the human body, three things are needed: fuelling, a thermostat, and so-called "effectors," which influence the generation and distribution of heat and the cooling in such a way that the temperature prescribed by the thermostat is maintained.

The body's fuelling system consists of trillions of tiny burners; that is, the cells, with the combustion heat of their metabolism. The effectors too are easy to find: stimulation or inhibition of the metabolism; intensifying or lessening of breathing activity; regulating the blood supply – the distribu-

tion of heat; making the body shiver with cold, which produces extra warmth by muscle movement; causing perspiration, which disposes of surplus heat by evaporation;
and other similar measures.

Up to 1960 the great problem was the thermostat. This
control device needs one or several thermometers, and a
control centre which compares the desired temperature with
the ones actually being recorded, giving instructions to the
effectors on the basis of any deviations. The control centre
has, in fact, been known to scientists for some time: it
consists of the nerve centre in the hypothalamus. This is a
region of the brain from which many other processes in the
body are controlled as well – processes accessible neither to
consciousness nor to the will. The heat-control centre,
therefore, is very close to all those command posts that it
instructs on how to regulate the values in the blood vessels
and all the other processes.

The only thing in doubt was where the thermometer of the
thermostat was situated. This was due to a completely
mistaken idea of what sensory perception really is. The
argument went roughly as follows: these thermometers
can only be temperature-sensitive sensory cells. But since
we feel heat and cold only on the skin, not in the interior of
the body, there are temperature senses only on the surface
of the body. Therefore the heat-control centre must somehow
be governed by the skin temperature. It does not even sound
probable, yet it was believed for decades.

The mistake was to think that, if sensory cells receive
stimuli, we should also be conscious of them in every case.
For there could surely be temperature senses in the interior
of our bodies which send the results of their measurements
to the heat-regulation centre without our consciousness
knowing anything about it. Perhaps there are sense impressions we never become aware of, which are yet of vital
importance for our life.

The problem was solved conclusively by the experiments
in 1960 of Dr. T. H. Benzinger,[13] Director of the American
Naval Research Institute, Bethesda, Maryland. Dr. Benzinger had himself locked up in an insulated steel chamber,
in which about a thousand instruments registered heat

radiation, heat flow, and perspiration all over the skin. His body fairly bristled with electrical thermometers: in the mouth, as radio probe in the stomach, in the rectum, in the auditory canal with light pressure against the tympanic membrane. He even had two thermometers surgically introduced (under local anaesthesia) into the frontal sinus just below the brain and deep into the nasopharyngeal space.

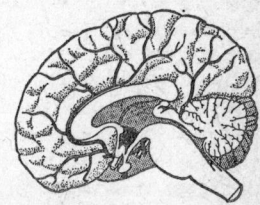

Figure 39. Man's heat-control centre is in the front part of the hypothalamus (black area). Nerve cells sensitive to heat measure here the temperature of the main arteries of the head, which run directly past them.

In hundreds of perilous experiments on himself, Benzinger discovered a place where the temperature state had a controlling influence on the heat reactions of all other places in the body. This could only be the thermometer of the thermostat, which had so long been looked for in vain. It is right by the heat-control centre, also in the hypothalamus, roughly at the centre of the head between the ears. Here the thermometer is closely surrounded by the two main arteries which supply the entire brain with blood.

So the thermometer immediately reports the temperature of the blood flow in the brain to the nearby control centre. Here the values reported are compared to the ideal value, which in the daily rhythm fluctuates a little around 98·6° F. (37° C.). If the blood temperature is only 0·018° F. (0·01° C.) above the ideal value, the control centre at once sends its instructions along nerve channels and hormone chains, in order to dissipate more heat from the body to the outside by activating the sweat glands or widening the small blood vessels. The latter accounts for someone's having a red face when he is hot.

If the blood 0·018° F. (0·01° C.) too cool, the body's power stations, particularly in the muscle cells, are "stoked up" to produce more interior heat. But with progressive degrees of heat or frost there is a limit. If this limit is exceeded, the human body's control breaks down. Man freezes to death or dies from heatstroke.

Curiously enough, we can play tricks on our interior thermometer, for example, by having an ice-cold drink on a hot summer's day. As the thermometer is also very near the pharyngeal cavity, it then measures a temperature below that of the blood. The thermostat, of course, misinterprets this and gives the whole body a terrific heating up. That is why the drink may have a brief refreshing effect, but leave one soon afterwards feeling hotter than ever.

So although man's consciousness has an inferior temperature sense, he does possess an absolute temperature sense, after all, in the thermometer of his thermostat. It is extraordinarily sensitive and exact, very much a match for the ultra-refined senses of fish and blood-sucking insects.

The thermometer is essentially a closely packed collection of heat and cold receptors. This was found out on dogs by three scientists of the John B. Pierce Foundation Laboratory, New Haven, Connecticut,[14] in 1964. They recorded from individual sensory cells by microelectrodes for listening in, warming the brain region of the hypothalamus at the same time by the insertion of miniature heating elements. Eighty per cent of the cells responded to the warming up by quicker firing of impulses and to cooling down by a slower sequence of signals; 20 per cent of the cells reacted the other way round.

But the thermometer at the main arteries is by no means the only temperature receptor of the heat-control centre. In any case human central heating would scarcely make do with the readings from one single "room." Further thermometers in the body's interior were discovered in 1964 by Professor Rudolf Thauer of the Max-Planck-Institute at Bad Nauheim[15], and this is how he did it.

He inserted a rubber balloon irrigated by cold water into the oesophagus and the stomach of anaesthetized dogs. They at once began shivering with cold as well as exhibiting other measures by which the organism tries to keep its temperature constant. This was proof enough.

The discovery was followed, as discoveries usually are, by attempts to apply, develop, and manipulate it, sometimes in bizarre ways. Experiments have already been started which aim to change the optimal value in the thermostat. A patient's body could be induced to reduce its temperature, facilitating the performance of difficult surgical operations.

Strange things happen to man's senses if his temperature has to be greatly reduced before an operation.[16] At 94·1°F. (34·5° C.) he can no longer see or hear. Below 85·1° F. (29·5° C.) the thermostat mechanism breaks down. The pupils open wide. Pain and any other sensations leave us. The pulse slows down to 40 beats a minute, and blood pressure goes down rapidly. At 80·6° F. (27° C.) respiratory activity ceases. The patient's state is like hibernation.

Dr. Benzinger, indeed, has the declared aim of putting man into artificial hibernation. At the Naval Research Institute, of which he is director, it is believed that manned space travel to remote planets will make it necessary to "switch off" the astronauts from time to time, like a technical device, during years of "uninteresting" flight, in order to save weight for provisions.

The Adaptation of Cold-Blooded Creatures to Their Environment

Ants would probably continue their hibernation all through the summer but for the messengers of spring which warm them up, releasing them from their sleep.

For the cold season they dig themselves a large hibernation chamber about five feet deep, which is sure to stay frost-free. There the ants, huddled together in their hundreds of thousands, become rigid with cold. They might stay in their cooling cabinet for ever, since the warmth penetrates only very slowly to the motionless insects in their cave. That is why they have "invented" a wonderful system for bringing the warmth of spring down to them. The French entomologist Professor Rémy Chauvin[17] gives a description of this:

Some ants are less susceptible to low temperatures than the majority. They stay lively, and all through the winter climb up from the hibernation chamber now and then to the top of the ant-hill. As soon as the weather gets warmer and the sun is shining, they even warm themselves in the open. When they return afterwards to their fellows, they give them a share of their warmth. Thus the temperature rises a bit, though very little, inducing other worker ants as well to "go and get" some warmth.

This is the beginning of a chain reaction which is made possible by the different sensitivity of individual ants to the cold. The more brightly the sun shines, the more warmth the spring messengers bring down, and the more ants "thaw out" and take part in collecting warmth. Finally, the great mass of the colony also begin to stir and return to new activity in the upper galleries.

Such is an early attempt at an air-conditioning system, which has been perfected to a high degree by other insect colonies. Von Frisch writes[18]:

> The constant temperature of 95° F. (35° C.) maintained in the breeding-cells is achieved by an amazing process. Workers crowd together in their thousands on top of the cells so as to make use of their collective warmth. In cooler weather they crowd together and cover the brood cells with their bodies, as with an eiderdown. On warmer days they scatter and, if the heat becomes excessive, they bring in water – as they cannot sweat – and cover the combs with a fine film which they cause to evaporate by fanning with their wings. They sit like little ventilators over the cells, driving the warm air towards each other and pushing it out again through the entrance.

In the winter they keep up an even temperature of 64·4° F. (18° C.) in their hive, even if there is a biting frost of 40° below zero F. (minus 40° C.) outside.

So it is no exaggeration to say that the individual bee is a cold-blooded creature. But in the community of the colony it has developed by well-organized team-work into a warm-blooded "super-organism". Even more striking is the analogy between the systems of man's central heating and the air-conditioning of termites. Professor Martin Lüscher,[19, 20] of Berne, Switzerland, has investigated this on the Ivory Coast of Africa with the indigenous species of termite, *Macrotermes natalensis*.

These termites need a tropical hothouse climate of 86° F. (30° C.) to live in. The two million termites inhabiting a medium-sized colony do not by any means rely on the natural tropical heat in order to maintain an even temperature, for this heat fluctuates a good deal. So they make themselves thermally self-sufficient. Just as in the human body every one of the innumerable cells supplies a little bit of warmth, so here each individual termite is like a small walking stove.

This private climate is protected from the temperature fluctuations of the outside world by walls hard as concrete and more than a foot and a half thick. Yet the two million insects must also be able to breathe in their fortress which has neither windows nor doors. They need over 260 gallons of fresh air a day. How is this provided?

By the artifice of an air-conditioning plant! On the outside of the termite hill several yards high, one can make out a bare dozen ridges leading upward. These are the radiator fins of the plant which the termites "invented" millions of years before man. Each of these ridges contains about ten narrow air vents leading from top to bottom just under the outer shell.

Figure 40. The air-conditioning plant of a termite hill. The radiator fins visible from outside are lined with ventilation shafts through which the air passes down into the vaults. From here the whole nest is ventilated.

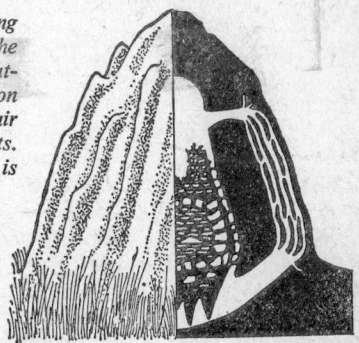

From the roof chamber where the hot, stale air collects, it floats into the air vents from above, where it cools off, at the same time making contact with the outer air through microscopically small pores in the stonework, giving off carbon dioxide and taking in fresh oxygen. From these "lungs of the termite hill" the air that has been renewed and brought to the correct temperature floats into the spacious cellar vault, which lies about three feet under the surface of the ground. From here the air circulates upward through all the rooms of the "apartment house."

There are about a hundred air vents within the radiator fins, and in them termite engineers are busy all the time, narrowing air vents down to a kind of valve, or widening them, closing or opening vents, depending on the time of day or year, on whether the temperature is too cold or too warm, and whether there is too much or too little oxygen.

Surprisingly enough, the termites regulate the ventilation so that the temperature ideal for their living conditions is always right at the centre of their nest, that is, in the queen's chamber. Who or what is it that at every moment informs the engineers about the climatic situation at the life centre of the state, so that they know whether they ought to turn the ventilation up or down? What substitute for a nervous system is there in a termite hill?

Is the news brought by messengers? With the enormous distances in the maze, their errand would take them several hours. Have the engineers themselves a thermostat set at the normal temperature? This seems implausible. Does the queen signal the temperature with special scents? This would be a singular phenomenon in nature.

The termites have such a wide scope for possible variations in their refined ventilation techniques that they could adjust to all conceivable environmental conditions in the whole of Africa, and could also exist quite well under Central European climatic conditions. One circumstance alone has spared Germany from being overrun by these creatures: with *Macrotermes natalensis* new colonies can be founded only by swarming sexual termites and in the initial stage, when king and queen, whose newly founded chamber lacks an air-conditioning system, have to depend entirely on themselves, they cannot survive in the German climate.

Life is much harder for other cold-blooded creatures, such as solitary insects, amphibians, and reptiles.[21] They can develop their vitality to the full only at a strictly limited optimal temperature. Any degrees below or above that slow down their motion progressively until they grow rigid with cold or heat, or even die of one or the other.

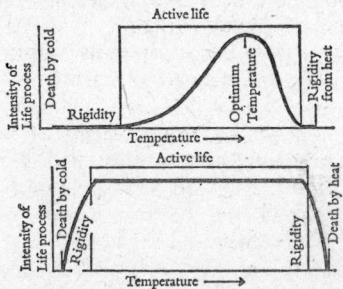

Figure 41. With cold-blooded animals (top graph) active life can develop fully only within a much smaller temperature range than with warm-blooded creatures and man (bottom graph).

Life itself is certainly a strange thing. In the universe the surface temperatures of the stars fluctuate from between minus 523·4; F. (minus 273° C.) to about 36,000° F. (about 20,000° C.). If we imagine this spectrum on a thermometer scale of eight inches, the only part of it where life can go on would take up a small fraction of an inch. Everything else is ice, desert, steam, and fire. The close-ups of the crater landscape on Mars have shown us in a shattering fashion how lonely living man is in the universe.

On this large scale the cold-blooded creatures take up only one or two tenths of that fraction of an inch. A fall of temperature by only 18° F. (10° C.) below their optimal life conditions slows down all their chemical life reactions to half, if not to a quarter. But what strange reactions these are! At 68° F. (20° C.) they make a desert scorpion go rigid with cold, and at the same temperature will kill a snow-flea. In inhabitants of the arctic the reactions occur at low temperatures with the same speed as those of tropical creatures at much higher temperatures.

It was once thought that the freezing point of water, 32°F. or 0° C., was the low limit of life, since metabolic processes in the body fluids must then also come to a stop. But the fish in the Hebron Fiord of North Labrador, already mentioned above (p. 85), still seem very lively at a body temperature of 30·04° F. (minus 1·7° C.) – that is, a little below the limit of life. This is not adequately explained by the hypothesis that they possess a sort of antifreeze in their body fluid. That is why Professor Herbert Precht[22] at Keil University studied these processes more closely.

The impression he got was roughly as follows: the "chemical factories" in the living cells which erect the edifices of the body from nutritive substances; the "power stations" producing energy for movement and chemical process; the "information links" sending signals; and the "hormone breweries" – all these apparently exist in our world in different designs. There are types which resist tropical conditions and others which are cold-resistant. For each animal species there is a different, specific design available, functioning best at exactly the working temperature which is the average temperature prevailing in that animal's habitat.

Up to a certain degree all these chemical plants which process molecules can even adapt slowly by themselves to fluctuating temperatures. A crab that has passed from a warm environment into a cold one, and at first lies almost inert, will gradually become acclimatized The chemical factories in its cells work faster and faster, until after a few days the initial convenient pace of work is reached again.

Chapter Three

Pain

The Sensation of Pain

AN American boxing champion, who shall be nameless, once had his nose flattened during a fight. He went on fighting unperturbed, as if nothing had happened. Later on he said that in the ring he had not noticed the injury at all; it was only after the fight that the sudden onset of pain made him aware that his face was disfigured and bleeding badly. Yet with dentists he had a reputation for screaming before the drill had even touched his tooth.

Surgeons from military hospitals and casualty wards also have surprising things to report about pain. During the Second World War the anaesthetist Professor Henry K. Beecher[1] of Harvard Medical School was surprised to find that two out of three patients with serious injuries felt very little if any pain on being admitted to a military hospital, although they were in just as bad a state as the relatively few who were screaming. Many refused painkillers or even claimed to have no pain at all. These soldiers were neither in a state of shock, nor had they become insensitive to pain, for they would flinch away at the mere jab of a needle.

Again and again the story was the same. All soldiers who were frightened of the war, who felt primitive fear and horror at the inferno of a battle to the death, with thousands of corpses gutted, burnt, trampled under foot – if such soldiers got wounded, it meant liberation from hell. The relief at having this time got away alive made them feel grateful for what had happened to them. Sometimes, in a rather macabre way, they were positively light-headed, a state of intoxication which suppressed their awareness of pain. The ones who screamed, on the other hand, were as a rule those who had always acted very tough beforehand, deceiving themselves about the reality of these horrors. When their illusions were abruptly shattered, they went to pieces.

Apparently there are more of the latter type in road accidents than on the battlefield. Doctors in casualty wards have found that four out of five seriously injured patients are so sensitive to pain that they have to be sedated with morphine. This suggests that they may have been the daredevils of the accelerator. Enveloped in a dream of lovingly tended paint and chromium, bedded softly in warm cushions, victim of the illusion of being a virtuoso master of tremendous horse-power, the driver may suddenly find himself in rags in the top of a tree by the roadside, or crushed between mangled pieces of metal. The mental crash may cause a good deal worse pain than the physical one.

It is now established that the amount of pain does not depend at all on the extent of the wound. The pain is anything but a predictable response to a stimulus that does us harm. Rather, what we feel is strongly modified by psychological influences, such as our experience, expectations, and – more subtly – our education, habits, manners, and culture.

Here is a striking illustration. There are some Indian tribes living along the Amazon where a woman breaks off work in the fields for only two or three hours to give birth to a child. Meanwhile her husband has for days been lying at home in a hammock, tossing about and groaning as if in great pain. Even after the birth, when the woman is back at her hard work, the husband remains prostrate, with the baby, to recuperate from his "terrible agony."

It sounds ridiculous, but religious rites, demon worship, and suggestion have produced in these primitive people ideas very different from ours. These Indians are firmly convinced that it is the man who "has" the children, even if in a higher sense (taken over from the world of the spirits), while the woman merely attends to the physical part of it, like a hen laying an egg. That is why the husband suffers real pain, and why the wife makes so little of the birth.

Our notion of labour pains is undoubtedly closer to the truth. But it may not be as realistic as we think. Did anyone ever see an animal in the wild that screamed when giving birth – and gave itself away to its enemies? And how about us? Hasn't it been possible for women to reduce their pain through training in natural childbirth?

Various animal experiments give us further insight into the phenomenon of pain. If it depends so largely on thinking about it, how do animals behave which have never in their lives had painful sensations? This question was asked by the Canadian psychologist Professor Ronald Melzack,[2] of McGill University. He brought up a number of Scotch terriers from birth in sensory isolation, protecting them to such an extent that none of them ever felt a single scratch, knock, or blow. Pain simply did not exist for them.

Later, when the dogs were grown up, they were abruptly exposed to brutal realities. When they bit each other in play, they seemed to feel as little pain as the boxer in the ring. They did not even flinch away or try to defend themselves, although wounds gradually formed and drew blood. To a burning candle they responded with curiosity and sniffed the flame. Of course they burned their muzzles badly. Even so, they did not learn their lesson, but kept on calmly sniffing any other open fire they saw.

This phenomenon must by no means be confused with so-called insensitivity to pain. Just as there are people who are blind or deaf, so there are children born who are numb to pain. Some fault in the pain conductors in their nervous system makes them absolutely insensitive to it. Such people have to cope with serious wounds, burns, and bruises in their childhood. Often they bite their tongues very badly when eating, and have extraordinary difficulties in learning to avoid injuries. There is no better evidence of the importance for our body of feeling pain than the fate of these unfortunates: for them the blessing, as we should at first imagine it, of freedom from pain has turned into a life-long scourge.

The nervous systems of the Scotch terriers, however, was

perfectly intact. They seemed to feel something, but could not make anything of it. They were puzzled, yet felt no discomfort. Like an Eskimo who for the first time experiences the tumult of a large town and cannot make anything of car engines "revving up", horns, police sirens, screeching brakes, or people shouting, although he can hear them all right – so to the dogs their first pain was a completely unfamiliar experience. Strange as it sounds, even pain is a thing which to some extent has to be learnt. This accounts for the enormous flexibility of the pain sense – from the Indians on the Amazon to the wounded soldiers.

Successful attempts have been made at combining sensations of pain with feelings of pleasure. If a strong electric shock is applied to a dog's paw, he usually reacts savagely, runs furiously round his cage, barking and growling. But the Russian Nobel Prize winner I. P. Pavlov offered dogs some tasty food immediately after each electric shock, and after many repetitions their reaction slowly turned from fury to delight. Eventually they would wag their tails directly after a shock and rush to the table where the food was. Their pleasure-stimulated behaviour was even maintained when Pavlov increased the strength of the shock to the utmost that could be borne, besides enhancing the effect by inflicting injuries and burns.

Something very dangerous is demonstrated here: the way emotions can be manipulated into an abnormal and absurd state through a conditioned reflex, as Pavlov called the transfer of a reflex release from a natural stimulus to one normally senseless. Since Pavlov, however, much subtler methods have been worked out by which even human beings can be induced to regard pain as amusing, wrong as right, and ugly things as beautiful. Often without those affected realizing it, the laws of conditioned reflexes are used to train members of special military units, for instance, to become masochists. As with the Japanese Kamikaze pilots who made suicide attacks on enemy ships, even the elementary urge for self-preservation can in this way be turned into its opposiite.

Still, surgeons have discovered a positive side to the way in which pain can be manipulated. When a patient complains so much of his pain that he has to have morphine injections every day, the doctors will eventually give him injections which contain nothing but a watery solution of sodium chloride. Thirty-five per cent of the patients treated thus find their pain alleviated as much as if the morphine injections had been continued. Since morphine, even in large doses, relieves acute pain only in 75 per cent of all patients, one may conclude that almost half the effect of the drug is illusory. In any case, morphine characteristically shows an effect only if the patient's fear level is high.

The Physical Model of Psychological Influences

If we have bruised a finger, many pain nerves will send a coded message about this event to the brain. But on their way the signals have to pass several "telephone exchanges" in the spinal cord and brain stem. Connected to these exchanges, however, there are also several "wires" which come from the cerebral cortex; that is, from areas where higher mental activity originates. These intervene by signalling a report on the state of the feelings, and can therefore either stop, amplify, or modify the transmission of the pain signals. These processes are so interesting, considering the phenomena I have just described, that it is worth having a look at the details which have so far been investigated.

Let us assume, then, that the bruised area of the finger is a square centimetre; several thousands of pain-sensitive nerve ends would translate the event into a code of nerve signals and pass these on probably over about two hundred conductors.

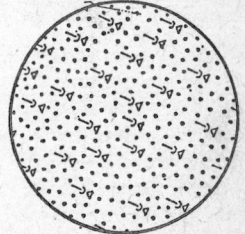

Figure 42. In one square centimetre of skin (here seen through the magnifying glass) man possesses about two hundred pain-sensitive points (black dots) and only twenty-five points sensitive to pressure (triangles), about as many as small hairs (dashes in semi circles).

How this transformation takes place is so far not known. Presumably in the destruction of cell tissues substances are released which, like scent or taste substances, have a chemical effect. This conclusion has been drawn from the fact that (in contrast to the sense of touch) between the injury and the sensation of pain there is always a brief latent period before, apparently, the substances take effect. But for the present this is no more than a hypothesis.

The pain receptors are a widely ramified rootlike system pervading the lower part of the surface layer of the skin.

The old idea that a pain-sensitive nerve ending is identical with a pain spot, which can be localized by a pinprick, has

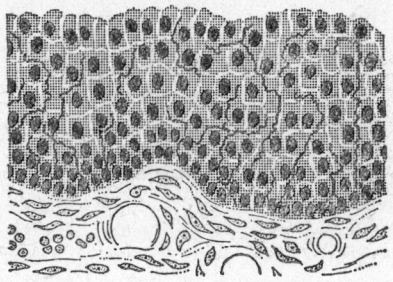

Figure 43. Cross section through the skin shows the cells of the epidermis (dark) and a network of pain fibres between the cells.

not been confirmed. Apparently everything is much more complicated. Perhaps there are zones of pain receptors, which have to be imagined as something like the receptor fields of the visual nerves (see p. 23).

From the finger the two hundred pain nerves run directly to the spinal cord. What arrives there is a pattern of signals at different speeds which already seems to us completely chaotic. We apparently have here a still unprocessed report of where what hurts, and how.

But at the next switchboard of the spinal cord, in an internuncial neuron or intermediate message transmitter, the facts are already being distorted. Professor Patrick D. Wall,[3] of the Massachusetts Institute of Technology, introduced microelectrodes in some of these neurons in cats and made a surprising discovery. The wires interfering here do not yet come from the brain but from the aching region of the body. Probably, however, they are not conductors of pain signals, but transmitters of touch sensations.

When Professor Wall irritated a place on a cat's paw by a

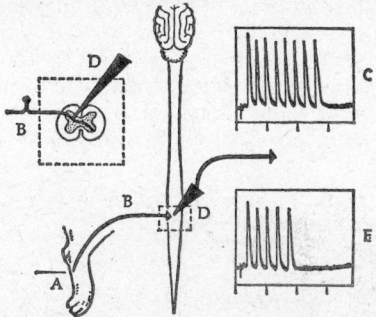

Figure 44. A pain nerve (B) signals a pin-prick into a cat's paw (A) by eight nerve impulses (C) registered by an electrode recording from the spinal cord. But if the point of stimulation is treated simultaneously with a massage vibrator, the pain nerve sends out only five impulses (E). This means that the pain has been reduced. Upper left shows cross section of spinal cord and indicating position of electrode (D) and pain nerve (B).

pin-prick or a painful electric shock, the listening-in oscilloscope from the spinal cord would show the excitation pattern expected. But if at the same time that he made the pin-prick, he treated its immediate neighbourhood with a massage vibrator, the pain report in the spinal cord was reduced to almost half.

This discovery led at once to a practical application. For it turned out that pin-pricks and electric shocks are no longer felt as painful by man either, if the skin at the place of the injury is massaged by vibration. So in future, injection and blood-transfusion needles as well as other sharp instruments can be introduced into the patient's body with the pain reduced by half.

In the spinal cord the pain signals on their way to the

brain pass other similar switchboards. At these points the cerebellum and the cerebrum already have an important contribution to make. Figure 45 shows how a pain signal (*a*) can be blotted out by an electric impulse (*a* + *b*) which the experimenter has sent through a nerve of the cat's cerebellum. Artificial electric impulses coming from the cerebellum have a similar effect (*a* + *c*).

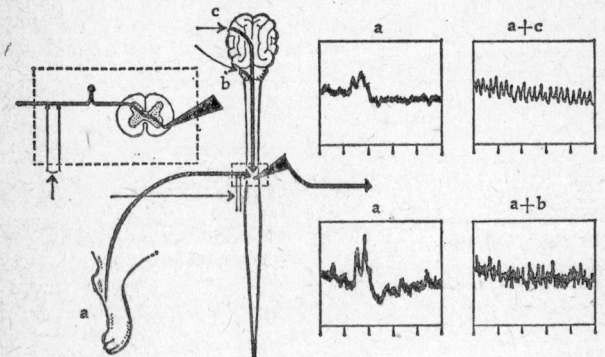

Figure 45. Pain signals from the cat's paw (a) can be completely blotted out by electric impulses from the cerebellum (b) and the cerebrum (c) (a + b and a + c). The a-graph here looks different from the recording shown in Figure 44 because the microelectrode was not inserted at the point of entry of the pain nerves into the spinal cord, but into deeper layers, and the nerves in the meantime have passed several relay stations.

Immediately before reaching the brain, the bundles of combined pain nerves from all parts of the body arrive at a giant telephone exchange with millions and millions of switchboards, in the so-called spino-thalamic tract.

Here there are such an enormous untraceable number of transverse connections that they seem to us like a confused mass of tendrils. Yet there is in everything a wonderful order which is beyond our grasp. Pain signals radiate from here into the most varied regions of the brain, producing emotions, memories, ideas, and prejudices. Thus varied and coloured, the impressions re-affect the spino-thalamic tract, modify the pain image, pass into the channels of individuality, habit, and custom, get lost in indifference, dullness, and apathy, or work themselves up into shock, hysteria, or hypochondria.

This psychological concoction is also the reason why a little harmless ache in a molar can dominate a person's whole being up to the very ends of his nerve fibres. The pain overwhelms the whole head, runs up and down the back, so that one's own diagnosis vacillates between inflammation of the sinus and injury to the spine. The pain humbles proud man, making a complete fool of him, while his great mind can think of nothing but just this hollow tooth.

A little is already known about the connections of the giant telephone exchange with the brain. In the spino-thalamic tract all circuits bundle together into five thick tracts, passing in this shape through the brain stem, and running into various regions of the brain.

So far Professor Ronald Melzack[4] has discovered the following about the significance of these five tracts: drugs that free man of the pain sensation without interfering with hearing or vision suppress completely the transmission of electric signals in tracts A, B, and C and reduce it in tract D, which, incidentally, plays an important part in the stimulation of alarm activities in the whole brain. Tract E, however, remains entirely unaffected by these drugs.

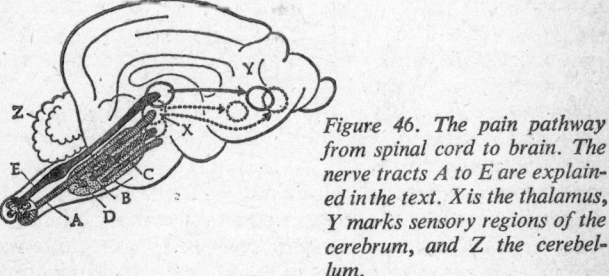

Figure 46. The pain pathway from spinal cord to brain. The nerve tracts A to E are explained in the text. X is the thalamus, Y marks sensory regions of the cerebrum, and Z the cerebellum.

Afterwards, when Professor Melzack severed the five tracts with his scalpel, this is what happened: the cut through tract A killed any sensation of pain in the animal; the same applied to tract C. On the other hand, cats with a severed E tract reacted quite normally to any stimulus, as before, while a cut through tract B had exactly the reverse effect: it made the animals hypersensitive. Some of them even showed spontaneous pain reactions although they had not been hurt anywhere.

These findings give us a rough idea of what normally happens here in the brain. Of course our body does not employ any such gross measures as anaesthetizing or severing of complete nerve cables; everything really takes place in a much more subtle and complicated way. I have already shown (on p. 21) how at a nerve switchboard signals from a group of nerve fibres can inhibit, prevent, or force the transmission of a message. The processes here must probably be imagined on similar lines.

Pain-Killing Operations

In his bed in the University Hospital, Chicago, Illinois, John Mulligan could no longer stand his pain. A growth below his left knee was driving him mad. The cramp in his calf had not relaxed for days, and his toes had become rigid in their contortions. In the end the twenty-seven-year-old man would have liked the leg that tormented him to be amputated. There was only one snag: the leg had already been amputated some years before.

Unfortunately, this "phantom pain" after the amputation of an arm or a leg is a quite frequent phenomenon. Although the limb itself is missing, the pain nerves leading to it are still intact. The scar tissue of the wound at the stump goes on functioning for many years after the operation. Slowly it contracts, and sometimes after days, sometimes only after years, it irritates the severed ends of the nerves, whereupon they send signals to the brain. The brain, however, cannot distinguish between these impulses and the genuine pain signals which used to come from the limb while it still existed. An impulse is an impulse; and so the oddest sensations may occur, like those caused by cramps, injuries, and growths which the patient perhaps never had in his life.

But even this is not the whole story. The irritation of the nerve ends at the stump is of course so minimal that the brain by rights ought to interpret it only as a slight touch. Yet it causes sensations of pain which in 30 per cent of all cases are very unpleasant and in 5 per cent actually give grounds for alarm. All signs point to the fact that the cause lies in a deep-seated confusion of nervous processes, brought about by the trauma of the operation; that is, in the encroachment of the psychological upon the physical field. This also provides an explanation of why operations to get rid of unbearable pain are so often unsuccessful.

Such operations have so far aimed at severing the pain nerves on their way to the brain. But the surgeon has been trying to hit only the nerves of the amputated limb and to spare the others; often, therefore, too little has been done rather than too much. The nerve fibres that remained were enough to reactivate the psychosomatic processes producing phantom pain.

In 1964 the brain surgeon Dr. R. C. Eggleton[5] tried a different method at the University Hospital in Chicago. He claimed that in the thalamus every pain-sensitive spot of the body has its specific projection point. If this tiny assembly of nerve cells amidst all the other pain assembly points from other parts of the body could successfully be eliminated, the patient ought to be rid of his torment. But a surgeon's scalpel is far too coarse for this operation; a sharply focused ultrasonic beam works much more delicately.

John Mulligan was the first patient on whom the new method was tried out. Dr. Eggleton bored a small hole in the skull of his patient, who was under a local anaesthetic, and through this hole he trained the ultrasonic beam exactly on the thalamic projection point of the phantom pain. The patient remained fully conscious and gave continual reports on his sensations. After every ultrasonic impulse the phantom pain diminished. After the seventh time it had disappeared.

The operative removal of other pains, for instance those caused by a cancer growth or a serious burn, is much more problematical. The surgeons Frank R. Ervin and Vernon H. Mark tried at the Massachusetts General Hospital, Boston, to repeat in man Ronald Melzack's method of killing pain in cats (see p. 105). In fact, man in this respect reacts exactly like the cat, and incurable patients were thereby given relief for their last days of life.

But this kind of pain-killing is irreversible. If the cancer patient recovers, the disconnected part of the body will stay insensitive all his life. Since 1965 American doctors have accordingly tried two other methods.

Professor Sean F. Mullan, Director of the Neurosurgical Clinic at University Hospital, Chicago, succeeded in putting the pain nerves to sleep for six months. He inserted a needle into the spinal cord through the small gap between the two top cervical vertebrae, applying first of all a very weak electrical probe of 0·3 milliampere. The patient then feels a tingling at some specific part of his body. As soon as the needle is moved by fractions of a millimetre, the tingling moves to other parts of the body, and one need only continue the procedure till the tingling has reached the aching spot: in this way the nerve responsible can be localized without fail. Then the current is increased to 1 milliampere and applied for 15 minutes; after which the pain is no longer felt. After six months the nerves regenerate (about the length of time in which the cause of the pain, for example a serious burn, will heal up). If they do not, the electrical chordotomy, as it is called, must be repeated.

The two Boston doctors, Ervin and Mark, have again led the way on the second method. A patient was in the last stages of cancer of the neck, suffering intense pain which drugs could no longer relieve. They introduced four silver electrodes through the skull into his brain, one each into cables C and D, the other two into the thalamus. From these electrodes four wires led into a little transistor. Directly the patient was in pain, he could press a knob and his torment ceased at once. The electric current, conducted into his own brain by artificial nerves, had stopped the natural pain signals.

One strange thing was that the patient, now free of pain, also stated that he felt as if he had had two Martinis; in fact he gave the impression of being a little tipsy. All this suggests that the current released other emotional reactions in his brain as well. Could this be the beginning of artificial control of our emotions, as by the Martian's antennae in science fiction? Instead of morphine, we press a knob. Instead of a bottle of spirits, we press a knob. The possibilities are too fantastic to go on.

The doctors saw an even weirder aspect of the phenomenon of pain when in their pain-killing operations they started to disrupt the nerve connections from the thalamus to the higher regions of the cerebrum, seat of conscious mental activity (see Figure 46). After such an operation patients reported that though they still felt the old pain, it was no longer disagreeable to them; indeed after some time they had completely forgotten the pain was still there. So the mere sensation of pain, on the one hand, and the obsession with it, on the other hand, of our entire thinking, feeling, and acting, are apparently two different things. This emerges, too, from the fact that the same patients were no longer worried after their operation about their illness being incurable, nor were they any longer afraid of death.

Can Insects Feel Pain?

One of the tall stories told by the great liar Baron Münchhausen was how his horse was cut in half by the portcullis of a fortress without his noticing this in the turmoil of battle. At the well in the market the animal drank and drank as if its thirst were unquenchable. At last Münchhausen tried to slap its rump, and his hand hit empty air. The water which the front part of the horse was drinking came out in a stream at the place where the rump was truncated.

Truth can be as strange as the Baron's fiction. Professor von Frisch, the great bee researcher, once saw a bee, which was sucking at a honey-pot, suddenly attacked from behind by a wolf-spider. It lost the whole back part of its body. The front half went on sucking in honey without pause (until it expired soon afterwards), though the honey poured out again at the severed waist – quite à la Münchhausen. Frisch later obtained the same results experimentally by severing a bee's hind part from its front part with a pair of scissors: the bee remained absorbed in its occupation and took no notice. Apparently it felt no pain. Its brain merely did not receive any longer the "I have had enough" signal which would have put a stop to its endless voracity.

These findings, and other examples of mutilation of insects, created the impression for a long time that arthropods were absolutely insensitive. But in 1965 Professor Vincent G. Dethier, a psychologist of the University of Pennsylvania, surprised the experts with the claim that insects too feel pain. It may even be the case that in some sense flies love, hate, suffer, and worry. The idea of them as dull little machines with a rigid inventory of unmodifiable, programmed instincts is false.

As evidence of his claim Professor Dethier offers a number of examples. The loss of a wing or a leg is certainly not ignored by the insect. Biochemical investigations have shown that the injured bodies at once shed hormones and other chemical substances into the bloodstream. Bees even seem to feel a sort of homesickness when picked up from a blossom and put in a cage. After only a short while chemical substances flow into their blood, sending them into a state of panic. If they are not set free, they die of a nervous disorder within a few hours.

Strong emotions were also revealed by a kind of lie detector. Since you can scarcely judge emotions from the look of insects behind their chitinous armour, Professor Dethier made recordings from various regions of their brains by microelectrodes and made a graph of the brain currents while the living creatures were being irritated. The results are still too confusing at present to form the basis of a theory; but they give further support for the idea of insects showing strong inner reactions which cannot be observed from outside. "When I look," said Professor Dethier, "at the stiffly armoured body, the rigid eyes and the silent movements, it seems to me a special challenge to find out whether there is not after all a real 'personality' behind them."

Should we call it love when an ichneumon-fly risks its life to defend its brood against enemies; hate, when hornets buzz furiously from a nest that has been poked; sadness, when an ant that has lost its way stays rooted to the spot? These are questions, of course, which cannot be answered even on the basis of hormone assays or graphs of brain currents; only the individual insects could tell us! But Professor Dethier's investigations certainly indicate that something "psychological" is happening inside them.

Fundamentally, we are faced here with the same phenomenon as with the human thermostat. We do not notice anything of our inner thermometers, yet they have a lasting effect on our body. Perhaps the insect is just as little conscious of its pain reactions. Or it may possibly register them, but without their producing any lasting agony, the same as with a person whose pain nerves have been severed between thalamus and cerebrum. Perhaps such agony would serve no purpose, since anyhow the insect cannot consciously take action to avert, say, the loss of a limb.

Besides, there are also human pain reactions which we do not notice at all. There are three different stages. First, there is the unconscious reflex which makes us wince when we have pricked ourselves. Secondly, we become aware of the feeling of pain, which forces us to do something about its cause. Thirdly, there is the mobilization of our inner defence mechanism: the "working teams" for repair of damaged bones and lacerated cell tissues, for defence against bacteria, special control of the blood supply, and many other such things. Of the extraordinary activity of the nerves and hormones which control these restitution activities, nothing penetrates into *our* consciousness either.

Strangely enough, this last aspect has proved particularly fascinating to electronics engineers. It gave Dr. William F. Hall[6] of the Northrop Corporation of Palos Verdes Estates, California, the idea of building electronic computers capable of fully automatic repair of any fault that might arise. At first sight this looks Utopian. But Dr. Hall contends that every living creature is among other things a mechanism created by nature with the magnificent faculty of repairing itself.

Pain is the alert at the occurrence of a defect. It produces many physiological measures for repairing the damaged part or replacing it with other organs. That is why the California team is at present investigating such natural functional systems, so as to obtain a pattern for electronic computers that will themselves notice, find, and correct their mistakes.

Such robots independent of human repair teams would be useful for automatic telephone offices, automated factories, and space stations, where faults have to be adjusted without human assistance.

Chapter Four

The Senses of Smell and Taste

The Marvels of a Dog's Nose

IN the dense undergrowth Ajax, the German sheep-dog, hit on the track of the "criminal". Sniffing, with his nose low to the ground, he ran left for about twenty yards. Then he turned, following the track in the opposite direction, and half an hour later hunted down the criminal in a thicket.

"Look," said the dog-breeder, who had followed Ajax on a long lead. "This is the best proof that the dog did not smell the track, but noticed it with some other sense so far unknown to us. For by the smell alone he could never have made out whether the track was running from right to left or the other way round."

This incident from the fifties faced Professor Walter Neuhaus[1, 2, 3] of the University of Erlangen with a difficult problem. As an expert on canine noses he had always thought it self-evident that dogs discover and follow tracks through their sense of smell. Was this not as obvious as was generally assumed?

But the experienced dog-breeder was ready for a second blow at science: "I can even put my rubber boots on. No scent can possibly get through those, yet the dog will find my track!"

An experiment bore out what he had said. But Professor Neuhaus set to work, starting all over again and going into the smallest details of the whole problem of finding tracks.

To anticipate, he managed to refute the breeder by means of new and surprising facts.

When a man walks barefoot over the ground, he loses about four-billionth parts of a gram of odorous sweat substance per step. This does not sound much. Yet it is an enormous lot if we count the odorous molecules clinging to every footprint: many millions of millions. Leather shoes prevent some of it getting through. But even here with every step some billions of butyric acid molecules are being pressed through – an amount which any tracking dog can still easily discover.

Rubber boots hold in even more, but by no means everything. The scent of the foot penetrates a brand-new rubber boot 0·2 millimetres thick within eight minutes. Rubber that is 2 millimetres thick absorbs odorous substances like a sponge within 38 hours. Although the human nose notices nothing of it, a dog still smells it easily. So here there is nothing mysterious involved after all, but only an extremely finely developed sense.

How is it, by the way, that man with his large protruding nose has a sense of smell so much inferior to a dog's? The answer is that his nose is first of all a preheater of air and a culture medium for a cold in the head, only to an amazingly minor degree an olfactory organ.

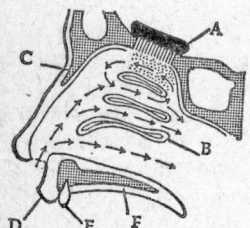

Figure 47. Olfactory cells fill only a small part (dotted area) of the mucous membrane in the "top floor" of the nasal cavity. Here the nerve fibres pass through the sphenoid bone (A). B = conchae, C = nasal bone, D = upper lip, E = incisor in the upper jaw (F).

Figure 47 gives an impressive picture of how remote from the ways of passage and how small the olfactory area proper is, lying deep in the interior of the nose, taking up a mere 5 square centimetres. The olfactory area of the German sheepdog's nose, however, measures 150 square centimetres. The disparity is even more blatant if we count the olfactory sensory cells.

THE SENSES OF SMELL AND TASTE

Man has	5 million
A dachshund	125 million
A fox terrier	147 million
A German sheep-dog	220 million

From a comparison of these figures one might jump to the conclusion that the German sheep-dog's sense of smell is 44 times better than man's. But this is not at all the case. Readings of the olfactometer have yielded the result that the dog's sense of smell is in fact one million times better. So the secret of the dog's nose lies not only in the enormous massing of the sensory cells, but above all in the way they function.

I will try to explain this by a fictitious example. An odorous substance which is sweated out by animals and human beings is butyric acid. A gram of this contains 7×10^{21} molecules – a figure at which the imagination boggles. If these molecules were made to evaporate evenly in all the rooms of a ten-storey office building – which, of course, is not a practical possibility! – then a man could only just register the smell if he stuck his nose into a door opening. But a dog would still react to the scent if the same gram of odorous substance were diluted to fill the air above the entire city of Hamburg up to a height of 300 feet.

With such a nose dogs are capable of extraordinary feats. Customs officers use specially trained animals for sniffing out toffee, tobacco, and opium. It is not even necessary to have the cases opened. One sniff at the lock is enough for the dog – even if the contraband is packed in tin, – for the tell-tale scent penetrates through microscopically small orifices. St. Bernards have become famous for rescuing people from avalanches: the human scent reaches them even through thick layers of snow and enables them to locate the victim's position.

In Holland and Denmark dogs are used for finding gas leaks. The animals are trained to stay put if they smell gas, and to bark at the place where the pipe is burst, although this may be a few yards underground beneath unbroken pavement. The work of these dogs is much more reliable than that of the most modern and sensitive apparatus for measuring odours.

A traveller in Africa told a most extraordinary story about

a dog's sense of smell. The traveller had booked in at a hotel and had his luggage taken up to his room, while he was having five or six whiskies at the bar. As he was about to go to his room, an enormous Newfoundland dog emerged from behind the counter and accompanied him to heel.

He could not remember his room number, or even which way to go for it. Was it here on the right? The alarming animal gave one bound to bar his way and started growling fiercely. Quite right – there hung a notice saying "Private."

So the man turned back. Oh yes, now he remembered – it was room number 122. He had scarcely opened the door when the dog pressed past him into the room, went up to his luggage, sniffed it, sniffed the traveller, then the cases once more, and without further ado trotted out of the room, wagging his tail.

One hardly dares imagine what would have happened if the scent of the luggage and of the man had not agreed. When the receptionist was asked whether there had ever been any thefts at the hotel, he said no, at least not since the dog had been on guard.

But how does a dog recognize the scent of a man as exactly as if it were a photo?

If a layman were asked to guess how many different scents there are in the world, he would probably say a few dozens, perhaps even a few hundred. A connoisseur of wine, who uses his nose to judge it, would count up the brands and vintages he knows, which would already amount to a much higher figure. A perfume manufacturer has worked out that a real expert must distinguish at least thirty thousand nuances of scent. But if a dog could talk, it would say: there are as many scents as there are men, dogs, cats, rabbits, deer, or other scented creatures all over the world, and many many more.

No two beings have the same scent, not even identical twins. Highly trained tracker dogs can also distinguish these from each other.

That is because animal scent is a mixture of a number of different-smelling fatty acids. With every creature this mixture differs according to the combination and concentration of the individual ingredients. Every person has a highly individual scent, unmistakably confined to himself, just as he has a face that is characteristic of him alone.

Now for the fact which the dog-breeder and Professor Neuhaus found so disconcerting: the individual scent ingredients in the smell of a track evaporate at different speeds. Within only a few seconds the "scent image" of a footprint changes so much that a dog can register this after running along a track for some twenty yards, and thereby make out the direction taken by the man, hare, or stag, he is pursuing.

By a trick Professor Neuhaus even managed to increase the smelling powers of dogs to three times the efficiency described so far. He fed his animals about one gram of butyric acid each. Two hours later the dogs' smelling powers had considerably deteriorated. But on the fourth or fifth day the animal could smell everything containing butyric acid three times better than at the beginning of the experiment.

For wild dogs and wolves this phenomenon is of great significance: four days after the dog or wolf has eaten something – that is, has also taken in butyric acid – he absolutely must have food again. It is just at this time that his optimally increased smelling powers help him to recognize and pursue even older, almost faded traces.

Food Attractants

In the twenties, a series of mysterious murders shook America. The murderer used to push his victims into a sack with some heavy stones and sink them in a lake. Even when the police had clues as to which lake had been used to dispose of a body, they were helpless. For how could it ever be found on the fissured bottom of turbid waters? And, according to the law, where there is no body, there is no murder.

Then an elderly Indian offered his services on condition that no one be allowed to watch him while he conjured up the dead. Each time it took him only a few hours to find the body. The experienced criminologists were amazed. Was there, after all, something in the magic the Indian claimed to have at his command?

At last the detectives found their competitor out. Here is

the account given by Karl P. Schmidt and Robert F. Inger.[4] The Indian had a large snapping turtle which he took to the lake. There he tied a long line to the reptile, letting it loose off a boat. Now he waited a good while and eventually had only to follow the line just as Theseus followed the thread of Ariadne. He caught up with the snapping turtle feeding on the body.

So the real detective was not the Indian, but the turtle with its amazingly good sense of smell. As the "dog of the American inland lakes," it noticed the corpse in the water and apparently headed for where the smell grew more concentrated, until it found its gruesome fodder.

Underwater a good nose works much better than eyes. Man, being a creature with a mainly visual orientation, finds this hard to imagine. Apart from vision, we are inclined to treat only the ear as being of nearly equal importance; and we regard everything else as mere accessories. Yet it is just the sense of smell, primitive and almost superfluous as it seems to us, which plays such a pre-eminent part in nature. Thousands of creatures might manage without eyes, but never without a nose.

This applies in the first place to fish, whose vision is much limited in the water. The development of blind fish in the eternal darkness of subterranean cave lakes is a proof of this.

The fishes' sense of smell is so essential to them that they can smell not only with their noses, but also apparently with large areas of their outer skin. In 1965 Dr. Mary Whitear[5] of the University of London investigated with an electron microscope the skin of minnows, those little fresh-water fish which often play in schools round the mooring places of our inland lakes. In the skin of the gill covers, the ventral region, and the tail, she discovered spindle-shaped nerve cells with which the creature is probably able to smell.

If man had the same faculty, he could tell, by merely feeling an object, which of his acquaintances had touched it in the course of the last day.

A regular war of smell is conducted by moray eels and octopuses in the coral reef. On this Dr. Irenäus Eibl-Eibesfeldt[6] of the Max-Planck-Institute for Behavioural Physiology at Seeweisen reports as follows: "In the dusk the moray eels come out of their haunts and steal upon their prey under cover of darkness. They take their bearings above all by their sense of smell. The defence tactics of the octopus, which are among the moray eels' victims, are adapted to this, In fleeing, they secrete a fluid, not in order to camouflage themselves as in the daytime, but to numb the predator's sense of smell for a while."

Incidentally, a man pursued by tracking dogs could produce a similar effect by pouring the urine of guinea-pigs on his tracks, should he happen to have this handy!

Some creatures can even smell through solid layers of earth or sand: the starfish is probably one of them. In 1961 Dr. S. L. Smith[7] observed the following incident on the Pacific coast of Washington: "A ten-inch starfish, creeping across the sand of the sea bottom, suddenly twitched slightly, returned to a spot it had just travelled over, and began digging. By careful and laborious work it made a hole 28 inches in diameter, which was 4 inches deep in the centre."

Anyone who has ever tried digging a hole in the sand with his hands under water, will be able to appreciate the achievement of the starfish. Boring in the wrong place would no doubt have unfortunate results for the creature's physical state. But it is a safe bet that it will find a clam precisely in the centre of the pit.

The suckers of the central disk cling fast to the shell of the clam. Then the starfish levers itself up on the arm points,

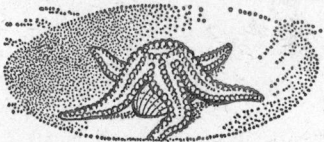

Figure 48. A starfish digs out a clam.

pulling the clam from the bottom like a cork from a bottle, to devour it in a very odd fashion. By the strength of its arms, the starfish pulls the two shells apart, inverting its stomach out of its body to cover the soft parts of its prey in order to digest them outside the body.

What Professor Niko Tinbergen of Oxford[8] reports about the smelling powers of the red fox sounds almost incredible. In the sand dunes of Cumberland he observed how this predator made such a large booty among hatching sea-gulls that it could eat only part of them. So it provided for leaner times by burying in the sand the birds it had killed and their eggs. But although the fox is supposed to be so cunning, it is evidently not cunning enough to remember where these caches are; so in later times of need it has to rely on its nose. It will exactly locate sea-gulls buried four inches deep in the ground when passing the place at a distance of three yards. Indeed it can smell out even eggs buried two inches deep if it happens to pass within a range of twenty inches of them.

Yet young rabbits, buried in the sand by their mother before she goes searching for food, are almost always missed by the passing fox. This is because baby rabbits have even less of a scent than eggs do.

Up to 1961 experts were still debating how the inhabitants of the soil, such as earthworms, wireworms, and larvae, find the roots of their host plants. The debate was as heated as most debates are where ignorance on the subject is nearly complete. The contention considered most likely was that these insects and worms burrow blindly in the ground, leaving it to chance whether they will find food or starve to death.

But nature never plays blind-man's-buff; and in 1961 Dr. J. Klinger[9] at last succeeded in proving that here too it is the "nose" which works minor miracles. For instance, the larvae of a sucking beetle, which feed on vine roots, have special olfactory senses which respond only to a single scent – that of carbon dioxide. This gas, which to us seems only very faintly acid, is excreted by the plant roots in comparatively large quantities and affects the larvae right through the soil almost like a "scent lighthouse."

Conversely, there are various plants which produce a scent that keeps leaf-eating insects away; and this may bear some relation to another problem which has proved difficult to solve: man's attraction for mosquitoes. Here too scent plays a decisive part.

It is well known, of course, that some people are seldom or never molested by mosquitoes, while others get covered with bites every night. This is sometimes said in jest to be due to sweet or sour blood, but that cannot really be the case. For the mosquitoes would have to bite first in order to taste the blood; whereas in fact they divert their flight at some distance from anyone at all uncongenial to them. So what either attracts or repels them must be some ingredient in the scent which everybody possesses in a different concentration.

If we could find out what makes the body uninviting to these pests, it might be possible to produce the ideal anti-mosquito pill. True, there are a number of anti-mosquito preparations: lotions, ointments, vitamin B^1 tablets. But they all lose a good deal of their effect as soon as one starts sweating.

We may envy the student discovered among a thousand test subjects by Professor Howard I. Maibach[10] of the University of California. Seventeen times this victim was exposed in the laboratory to a swarm of starved mosquitoes thirsting for blood; yet he was bitten only twice. Afterwards he was sent to a steam bath and kept there until he had produced three and a half pints of sweat.

While this book was being printed, biochemists examined the ingredients of this sweat, to compare them with the scale of odorous substances in those subjects who had been devoured by the mosquitoes in the laboratory. Perhaps the decisive substance can be found in this way. With its help science should succeed in freeing mankind from the scourge of mosquitoes.

An entirely fresh tack was given to these efforts when information became available in March 1968 of new experiments in Canada. The zoologist Dr. R. H. Wright, working for the British Columbia Research Council,[11] had been the first to explore the approach tactics of these bloodthirsty insects by means of a special wind tunnel. A stream of air of regulated temperature and humidity could be blown through this apparatus and mixed with appropriate odours. A trace of smoke was added to make the air movements visible. The approaching insect was also caught in a beam of light, so that it became clearly visible and its course could be followed photographically.

A slight rise in the carbon dioxide concentration of the air, such as would be caused by the breath of man or other warm-blooded creature, serves as a starting signal for the mosquito as it sits on a wall. At first the tiny predator will buzz around without much direction. It tries merely to fly against the stream of air.

This more or less random movement changes as soon as the mosquito runs into a zone of air that shows a relative increase in warmth and humidity. Here the insect reacts with amazing sensitivity. It is capable of sensing differences in temperature of only one five hundredth of a degree Centigrade (see p. 78). The breath of air can be so weak that the body's own thermal currents are completely sufficient. The body odours that had been sought so intensively by other scientists do not play a role up to this moment.

The buzzing, bloodthirsty little creature does not change its course when it encounters the warm and humid stream of air from its host. Now it is on the right track. But as soon as it becomes aware that it is leaving the warm and humid zone it begins to tack against the wind. These manoeuvres continue in a zigzag course until the mosquito approaches the un-clothed or only thinly covered part of a human body closely enough for it to come in for a landing under the guidance of its visual or tactual sense. It is therefore the warm and humid stream of air that guides the insect as precisely as a radio beacon brings an airplane down the approach pattern to the runway.

Odours can serve only as repellents. They change the insect's behaviour, so that it keeps moving away from, rather than approaching, its potential host. When the mosquito gets wind of the odour, which is "unpleasant" for it, its flight programme changes as follows: the resting insect starts to fly as soon as it notes an increase in the carbon dioxide content of the air or merely the repellant odour. It flies around, but contrary to the approach behaviour, avoids any breath of air that is warmer or more humid than the remaining air. Thus it put more and more distance between itself and the source of the unpleasant odour.

This behaviour led Dr. Wright to an important conclusion: if the presence of the repellent odour sets off the avoidance behaviour, regardless of its source, then it is quite unnecessary to spray or cover face, and hands with the usual insect repellent spray or solution, as one usually has to do in areas that are infested with mosquitoes. It is much more effective and pleasant, according to the scientist's opinion, to spray the bedroom floor or articles of clothing, preferably the socks, with the solution.

Now, for the first time, it is also possible to construct a mosquito trap. One has to imitate a body which gives off a warm, humid stream of air containing carbon dioxide, but in such a way that it becomes more attractive to the blood-thirsty little creature than a human body. Dr. Wright's solution is amazingly simple: a stovepipe, two feet in length, containing a burning candle (CO_2 + warmth) that heats a small container of water (humidity). The attractant air, warm, moist, and rich in carbon dioxide, seeps through a porous styrofoam top. After some time, all mosquitoes in the room land on top of the stovepipe and keep on biting it. If a contact poison is applied to the styrofoam top, the mosquito problem inside the room is solved.

The Divining Rod of the Ichneumon-Fly

In *1984* Orwell forgot one important point, Big Brother's prying nose. My description so far of animals' smelling powers will give some idea of what would happen if, by industrial espionage in the realm of nature, man artificially produced smelling apparatus capable of the same achievements as dogs, snapping turtles, or mosquitoes. Masked television eyes and hidden microphones would seem child's play compared to the spying methods then possible; but of course there would also be a great many ways of applying this and other discoveries for the benefit of humanity. This book, in fact, is not intended as a mere enumeration of curiosities, but to offer a glimpse into sense worlds that may one day be brought into our lives in a highly artificial way.

Rather as the dowser can detect hidden water, various species of ichneumon-flies can hunt out victims which are hidden deep in the wood. Female ichneumon-flies have the peculiarity of putting their eggs into or on other insects, such as caterpillars, spiders, queen ants. Some species, oddly enough, specialize on insect larvae which live an inch or two deep in the sound wood of tree trunks. Seen from the outside, there is not the slightest indication of this involuntary host, but the ichneumon-flies can find it.

The North American species *Megarhyssa lunator* runs excitedly to and fro, up and down the tree trunk. Suddenly it stops and creeps back a bit, improving its position a little, then pushes its ovipositor quickly down to a depth of three inches into the tree. In most cases it hits the hidden larva exactly.

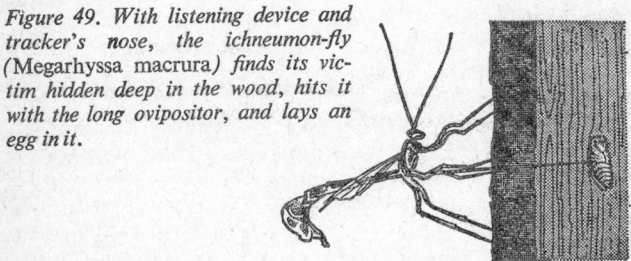

*Figure 49. With listening device and tracker's nose, the ichneumon-fly (*Megarhyssa macrura*) finds its victim hidden deep in the wood, hits it with the long ovipositor, and lays an egg in it.*

The first light on the hitherto mysterious detection methods of the ichneumon-fly was shed by observations of the American entomologist Professor Harold Heatwole[12, 13] of the University of Michigan. When a female of one of the three *Megarhyssa* species has completed her larva stage inside the wood and is working her way out, males (from all three species) with mating intentions have already taken up position at the place where she is expected to emerge; they are attracted by the chewing noise of the wood-crunching female. "When amplified by a loudspeaker," says Professor Heatwole, "it sounds like something like a man eating raw carrots."

The males sometimes make mistakes, however. Professor Heatwole observed waiting ichneumon suitors faced with a beetle breaking out of the wood, so they had to listen around for new drilling noises. But if a female does appear, the suitors have first to find out which of the three closely related species she belongs to. They feel her and smell her with their long antennae, and, after prolonged zoological evaluation, only the male of the identical species copulates. In view of this discovery it seems very likely that the female ichneumon-flies also locate their objective by the noise it makes in the wood, through their listening device, that is, acoustically. Their ears, incidentally, are placed on every one of their six feet and consist of cells sensitive to vibration.

But though the male suitors can afford to make mistakes, the females when they lay their eggs cannot. A female ichneumon-fly can safeguard the life of her descendents only if she lays her egg into the larvae of the horntail. If by mistake the egg got into a different insect living in the wood – for example, a larva of a long-horned beetle – the parasite would either be killed by the host's inner defence mechanism, or the host would die prematurely, taking his uninvited guest with him.

Accordingly, since errors are fatal for the ichneumon-fly, precautions have to be taken; in this case probably by the "nose." While the mother is creeping along a tree trunk, it smells the living creatures below. Perhaps it is alerted, by a specific tempting smell, only when above the right host creature for its eggs. The detection of the exact spot is then taken over by the listening device.

But the larvae seem to possess a defence against the combined operation of detection. Directly they hear the drilling noise of the ichneumon-fly's sting, they stop wriggling and keep dead still. Professor Heatwole established that the accuracy of aim is then considerably impaired.

Many ichneumon-flies have the bad luck to lay their eggs into host creatures which have already been injected with eggs by another ichneumon-fly. Nature's planning has not provided for this overpopulation of larvae, so it is usually fatal for all concerned. At best only one larva survives after a fierce competitive struggle. Evolution has therefore produced species of ichneumon-fly that avoid these mistakes. *Horogenes chrysostictos*,[14] which attacks the Mediterranean flour moth, can for instance distinguish healthy host creatures from those already occupied. Probably the species makes its diagnosis from the scent.

This discovery gave the physicians the idea, to which I shall return later, of diagnosing human diseases by an olfactory process.

Dr. R. L. Doutt[15] proved that the ichneumon-fly *Pimpla bicolor* has an extraordinarily keen sense of smell. This fly is parasitic on the cocoon of a South African moth. For his experiment Dr. Doutt had chosen a wood in which *Pimpla* ichneumon-flies had never been seen before. Here he opened up a single cocoon of the moth, which did not frequent the wood either. A puff of scent rose from the cocoon, and a quarter of an hour later a whole swarm of *Pimpla* females appeared, which settled not only on the cocoon but also on Dr. Doutt's scent-sprayed hands.

When we eat our daily bread, we could really pay a tribute to the fine nose of the chalcis, *Lariophagus distinguendus*, which keeps in check one of the most harmful insect species: the grain weevil. During the past decades this beetle has spread from the south of Europe all over the world. Three to four millimetres in size, it lives in stored grain, and to deposit an egg bores into a grain of either wheat or rye. The offspring then feeds till the grain is hollow, emerges as a grown beetle, then goes on eating its fill from the food that surrounds it.

The grain weevil would have a wonderful time if it were not for the chalcis. Dr. A. H. Kaschef[16] tested the capabilities of this grain-controller as follows: with a heap of 96,000 sound grains of wheat he mixed in 118 grains attacked by the grain weevil. All but four of these were picked out by the chalcis, although they differed from the sound ones only by the scent and a tiny hole, and were hidden in the heap to a depth of up to thirteen inches.

This could serve as a model for scent-controlled sorting machines of all kinds.

The Scents of Home

From the depths of the sea the fish rise to assembly into schools of hundreds, thousands, even millions. [Dr. Friedrich Dörbeck[17] describes thus the salmon migration in the East Siberian Amur River.] At first individual fish stay at calmer places, near the bank or in depressions in the river bed. Their number increases all the time. Now they venture into the centre of the river and suddenly they press forward in great masses. The river-bed is now overcrowded by their huge numbers. No obstacle can deter them. They jump over sand-banks several feet wide and tree trunks drifting in the river; lying on their sides wriggling and splashing, they roll over rapids. At deeper spots they stop for a short rest. These resting places are then so overcrowded that more salmon are to be seen than water.

Around 1900, in the northern area of the Far East alone, the annual number of salmon migrating from the sea into the rivers was estimated at 400 million. Since then the canning industry has taken to using water-wheels to dredge from the river the fish at the resting-places. Contamination of the rivers by sewage has done the rest, and so this magnificent natural spectacle is gradually becoming a rarity, especially in Europe.

Where do the salmon come from, and what are they after in their mad rush forward?

With some species, when the salmon plunges wildly upstream, it is seven years since it was hatched in a clear mountain stream, a tiny transparent creature less than an inch in size. (With other species the period is only two, three, or five years). In the first three to four months of their lives the new-born still feed from a sort of haversack their mother has endowed them with: reddish yolk sacs hanging under their bellies, destined to help them over the lean winter period.

Once this food is finished, the little salmon, now grown to a length of two inches, must work itself out of the hollow dug by its mother and go foraging in the running spring water together with thousands of its brothers and sisters. The river soon carries it into the sea, where it leaves the other salmon to follow its own course and embarks all alone on a global tour of thousands of miles.

Seven years later, with the punctuality of an overseas freighter service, the full-grown salmon reappears at the river mouth. Then a hectic marathon begins, a swimming contest with an army of rivals and with death.

In western Canadian rivers this gruelling race goes on over

a distance of a thousand miles upriver, as Dr. J. R. Brett[18] of the Canadian Fishery Research Institute in Nanaimo reports. With an average speed of two and a half and a top speed of four and a half miles per hour, the salmon drive themselves on almost without a pause, day and night, without feeding, towards their goal – for two to three weeks.

The enormous physical strain uses up 96 per cent of the body's fat and 53 per cent of its protein substance. The feeding and digestive organs atrophy almost into disappearance, while the organs of propagation grow tremendously. The salmon lose their silvery sheen and become first a dirty yellow and eventually an olive green. The male's mouth lengthens and curves into a shape like a pair of pincers. An ugly hump of inedible flesh grows on their necks. The gill plates become hard and brittle, and their respiratory powers dwindle more and more.

Despite their poor physical condition the male salmon fight each other fiercely as soon as they have reached the spawning places. In the meantime the females with springy movements dig hollows about ten inches deep in calmer stretches of the bank of the river-bed, and deposit about two thousand eggs in each. The males that were most successful in the contest now have the best chances of being chosen to inseminate the spawn. A short while later both parents die.

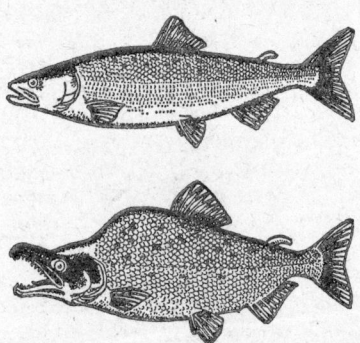

Figure 50. Salmon change their appearance in a horrifying way when they travel upstream. While the back of the female (above) becomes almost straight, the male's (below) changes into a hump.

Exciting as these facts from the salmon's life may be, the real sensation did not come till 1939, when Dr. W. A. Clemens,[19] a Canadian, carried out an amazing experiment. In a tributary of the Frazer River system he had marked with tags 469,326 young salmon heading for the Pacific. Years later he caught 10,958 returning salmon with this mark. The most astonishing thing was that in no other tributary was even a single marked salmon caught.

The fish, therefore, had not simply swum from the sea into just any river and from there into its next-best tributary in order to spawn at a convenient place. The terrible effort of the last few days of their lives had had only one goal: the place where they were born. Not a single salmon had lost its way in the search for its home.

Soon after this discovery the wildest speculations were current as to how the salmon manage this phenomenal feat of homing. There was talk about magnetism or a mystical sense for their first home. At the end of the fifties zoologists at last solved the problem.

In 1957 Dr. L. R. Donaldson and Dr. G. H. Allen[20] made the following experiment: from the water of a river in the Rocky Mountains they fished fresh salmon spawn and transported it several hundreds of miles into the mountain stream of another river system. What were the salmon now going to consider their home: the mountain river in which from time immemorial all their ancestors had been hatched, and died, or only the place which they had memorized as the starting-point of their journey in their earliest childhood when they were still tiny creatures?

The result was clear and unambiguous. After they had lived in the sea for years and were fully grown, the homecomers without exception made for the river into which the scientists had transported the spawn. This was the end of the theory about a mystical feeling for the first home. On the other hand, the salmon seemed to remember exactly the route which they had swum in the opposite direction years before as tiny young fish. This discovery is surely far more miraculous than the theory which has now been exploded.

But how can we possibly believe that a rational recognition of the river and its right tributary has taken place? We can hardly assume that the salmon retains an image of the land-scape above or below the water. With the flair of the enthusi-astic zoologist, Arthur D. Hasler[21] of the University of Wisconsin formulated the right working theory, which sup-posed that the salmon take their bearings from the scent of their native stream and follow a characteristic trace of scent upstream like dogs following the scent of a track.

That fish have a sense of smell has already been shown by the example of the minnows and moray eels (see p. 118). How highly developed the sense of smell can be in fish was demonstrated by Dr. Harald Teichmann[22] in 1957 from the example of the eels. The result of his experiments at Giessen University sounds almost incredible: if a thimbleful of syn-thetic rose scent were diluted by fifty-eight times the amount of water in Lake Constance, the eel could still recognize the scent. This surpasses even the smelling abilities of a highly qualified tracking dog.

But does the smell of rivers really differ? Professor Hasler proved that particles of the vegetation in the catchment area give the river its characteristic odour. In the laboratory he successfully trained salmon to react to water samples of various rivers.

Professor Hasler obtained the final confirmation of his thesis by the following experiment. In two arms of the Issa-quah River in the state of Washington, he caught 302 return-ing salmon and transported them downstream again below the fork in the river. There he blocked half of the fishes' nostrils with cotton wool stoppers and released them again. The salmon with unblocked noses unanimously chose the same tributary into which they had swum the first time. But the fish that could no longer smell had entirely lost their bearings. They kept tacking this way and that and could not make any definite choice of direction.

This suggested to Professor Hasler that he might develop a synthetic odourous substance with which fishermen and breeders would be able to lure the salmon into desirable tributaries and to selected spawning places.

For at present salmon will never again visit a spawning place where all the salmon fry have once been destroyed by predators, fishermen, poisonous effluents from factories, building of dams; and this applies even if the cause of destruction has been removed. You may catch salmon ready for spawning and take them to such a place, but it is no use. An irresistible instinct stops them from carrying out the act of propagation in water where the scent of their birthplace is missing. They die without depositing their spawn.

But if salmon fry in breeding places are imprinted with a suitable synthetic odorous substance, it should be possible through this guiding scent to recolonize abandoned waters. particularly for rivers in Germany, where salmon are almost extinct, such an operation would be of great value.

Love Scents

A female butterfly is all alone. The nearest male of her desire is waiting for a sign seven miles away. How can she at least make her presence known? It sounds a hopeless situation.

By calling? Impossible: no human voice, let alone a butterfly's, will carry that far. By optical signals? Impossible: the facet eye recognizes pictures only from a very short distance. By giving off a perfume? Impossible: even a highly qualified tracking dog can only follow a track; it cannot smell out a criminal directly at that distance.

Impossible – that is what man is always thinking when he judges others by himself. It is in fact quite possible; for the male of the silkworm does fly direct for many miles to find his female. There *are* "shouting" butterflies, emitting ultrasonic cries: I shall return to these later. And butterflies do smell each other; this has been proved by Nobel Prize winner Adolf Butenandt,[23] and his collaborators, Dr. R. Beckmann, Dr. Dankwart Stamm, and Dr. Erich Hecker.[24]

It is certainly rather fantastic. The butterfly female possesses hardly a ten-thousandth of a milligram of special perfume. At any given moment only a minute proportion is exuded into the air, where it is further rarefied to a point beyond our imagination. These inconceivably faint traces, at a distance of several miles, can rouse lively reactions in the males, which lead them to the source of the scent.

Apparantly only a few scent molecules from time to time need to hit the antennae of the male butterfly. This highly sensitive receptor, which looks like a palm frond, is virtually a part of the nervous system sticking out of the head. No less than forty thousand sensory nerve cells of various types are closely packed in together here in a tiny space.

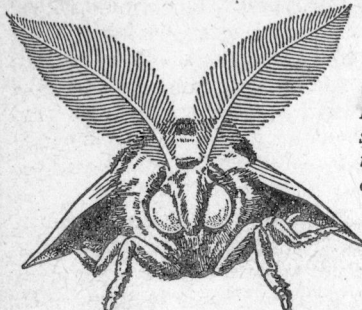

Figure 51. No less than 40,000 sensory nerve cells are packed together on the 'palm-fronds' of the silkworm moth.

But these few scent molecules exercise a dictatorial control over the silkworm moth during its brief span of life. This life begins when it slips out of its pupa. At first it hardly moves, does not fly at all, but remains inert, resting in a sheltered spot.[25]

It does so with good reason, for throughout its life it lives only on air and sex. In contrast to many other butterfly species it is incapable of absorbing even the tiniest morsel of food or a drop of drink. When the energy ration in the shape of fat which it took with it on leaving the cocoon is exhausted, it dies. To fulfil the function it was born for, it must husband its strength. Flying about at random in the hope of finding a female would be an unpardonable squandering of that strength.

So it sits and waits patiently until the wind blows a few female scent molecules at its antennae. Then it has no more choice. As if drawn by invisible strings, it soars aloft, beating its wings unceasingly until it either reaches its objective or its strength fails.

How does the butterfly, alerted by a few molecules, find the far distant female? What technique enables a small insect

THE SENSES OF SMELL AND TASTE   133

to achieve so much more than a dog, which always depends on its track?

There are two fundamentally different methods by which a creature may locate sources of scent. One is the gradual approach by sniffing its way through territories with a steadily increasing concentration of scent – that is, the following of a scent gradient. If the creature registers the tempting smell more strongly, say, in the left nostril or on the left antenna than on the right, it will keep left until the smell on the right has increased; it is thus drawn nearer and nearer its goal along a wavy course. This, for instance, is how bees wind their ways to blossoms, the scent of which they know. If their two antennae are bound crosswise, they fly away in the opposite direction to the source they are after.

To the silkworm moths this technique of location would not be much use. Some miles away from the female, where the sexual attraction is so much diluted that there are only a few individual molecules floating about here and there, their distribution is haphazard and there is no scent gradient. So nature has produced another invention: the scent merely stimulates the silkworm moth to flying activity, which is then controlled by two anemometers in the joints of the antennae head on to the wind. In this way the moths also reach their goal by flying upwind.

For the technique to be as efficient as possible, there is another essential condition. When a dog lifts its nose in a wood, some molecules of deer scent will no doubt get into its olfactory organ, and so will some molecules of hare, partridge, mouse, and human scent. But the characteristic scents of these creatures and of man are a mixture of dozens or even hundreds of different types of scent molecule; so at a long distance away these are unrecognizable fragments to the dog. It will not be able to do much with them, unless it gets into its nose molecules of all the types belonging to a scent.

So that it can be easily distinguished from much farther away, the sexual attractant of female butterflies consists of only one type of molecule. Conversely, among other olfactory sensory cells on the male antennae, it has some which react only to this one type of molecule and to no other odorous substance in the world. These cells represent, in a way, a special nose for a single scent. This is why the butterfly can smell much farther than the dog, although both creatures have a sense of smell which is roughly equal in sensitivity.

What applies to the silkworm moth goes also for innumerable other insects. The members of most other insect species also attract their partners by scent signals; other butterfly females, bumble-bee males which extract their attractant from blossom oils,[26] fruit-flies; also the walkers and crawlers like cockroaches, cucumber-beetles and grain weevils, which emit smell only at very special times of the night (one species from 11.00 p.m. to 4.00 a.m., another from 2.00 to 6.00 a.m.), so as not to reveal their presence to their daytime enemies. Even water-bugs give off scent.

Professor Martin Lindauer[27] estimates that there are about half a million different attractants in the insect world. Many of them we cannot register at all, but some to our noses smell of pineapple, chocolate, lemon oil, musk, flowers. An odorous substance released by a tropical water-bug smells of cinnamon. People in South-east Asia squeeze out the creatures' glands and use the scent for the preparation of their rice meals. In 1957 Professor Butenandt and his colleague Dr. Tamm succeeded in determining the structure of this so called Belostoma scent and then producing it synthetically. It is now being sold to Asia as a spice.

In almost every species of insect the molecule of the sexual attractant looks different.

The scent molecule is the key to mating between partners of the same species. If the species does not coincide, the creatures do not copulate, even if the structure of their bodies is nearly the same. Thus the shape of the molecules prevents hybridization between different species.

Here is an example: in the family of the *Drosophila* fruit-flies, the tiresome tiny creatures which crop up in great numbers anywhere that fruit is rotting or fermenting, there are 2,000 different species. Each species apparently has a somewhat differently constructed sexual attractant molecule, for the members of different species cannot tolerate each other's smell. But if the female's antennae are amputated, the barrier between the species is immediately gone. Members of different species forget that only a short while ago they could not stand each other, and without inhibition start producing hybrids.

However, there are exceptions to the rule. The scent of the female tobacco-worm moth, for instance, stimulates not only the male of its own species, but also the sexual needs of the male of the Indian flour-moth. This species comes fluttering

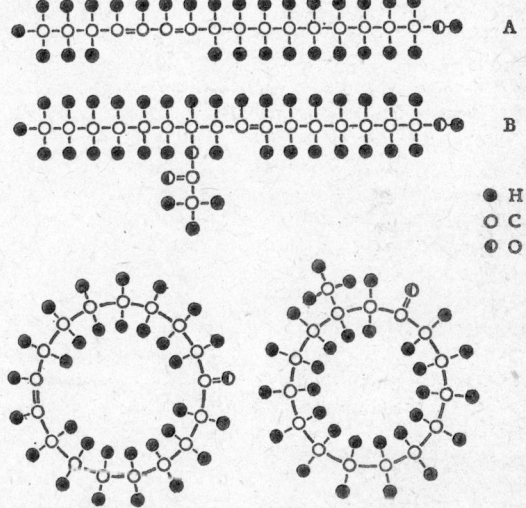

Figure 52. The structure of the attractant is amazingly similar yet still different for the silkworm moth (A) and the gypsy-moth (B), for the civet cat (C) and the musk-deer (D). H = hydrogen, C = carbon, O = oxygen.

along at once, attracted by the scent, and makes an attempt at copulation, although its efforts are doomed from the start for purely anatomical reasons. The flour-moth is dwarfed by the enormous female of the tobacco-worm moth. So here nature had accidentally created the same attractant twice.

The most sophisticated "scent deception" is practised by the South American jungle plant *Ophrys insectifera*. From its blossom the plant emanates a scent which is almost identical with the sexual attractant of an indigenous butterfly. More-

over, the shape of the blossom also has the essential characteristics of the female butterfly's body. The male promptly
falls for this trick of nature's, tries fervently to copulate with
the blossom, and thereby pollinates it.[28]

Besides their purely scientific value, research into attractants has great practical significance. It is tempting to reproduce the attractants of harmful insects synthetically and use
them to lure the creatures into death traps and eradication.
Following this course, however, produces its own problems.
Without the large-scale use of highly poisonous pesticides,
insects would not leave enough vegetable food for the steadily
growing world population. Thousands of millions are being
spent on chemicals in order to save agricultural products
worth ten times as much. But at the same time the insecticides
are beginning to turn into a danger also to birds, fish, game,
and human beings; while the pests gradually develop an
immunity to the chemical substances, we slowly poison
ourselves.

In the last resort only synthetic sexual attractants will be
able to save us from this dilemma. These substances are not
poisonous and are dangerous to only one species of insect.
Besides, comparatively small amounts are sufficient, and the
pests cannot develop any immunity to them.

Unfortunately, they are extremely difficult to get hold of.
Before they can be produced synthetically, their atomic
structure must be known. But how can you examine the
chemical structure of a substance, when you obtain only a
few milligrams of it by squeezing out the abdominal glands
of a half a million insects?

After twenty years' experimenting, extraordinary efforts
and many disappointments, two teams of scientists at last
managed this feat in 1959 by using the most recent technical
apparatus: a team headed by Professor Adolf Butenandt[29]
at the Max-Planck-Institute for Biochemistry in Munich,
succeeded in analysing the structural formula of the sexual
attractant of the silkworm moth and producing this substance
synthetically a short while after. Almost at the same time
Dr. Martin Jacobson and Dr. Morton Beroza[30] of the United
States Department of Agriculture Insect Research Institute in
Beltsville, Maryland, achieved the same with the sexual

attractant of the gypsy-moth, the caterpillars of which cause great damage to fruit and forest trees.

It is very surprising that the long-sought substances are no complicated compounds with mysterious chemical properties. They are merely simple hydrocarbon chains with a few side groups. As can be seen from Figure 52, both substances are very similar. In the synthesis, of course, the exact geometrical arrangement of the atoms within the chain molecule has also to be precisely reproduced. No butterfly can be attracted by a synthetic scent molecule with the same chemical formula, if there is a wrong spatial arrangement of even a single atom.

Since 1960 Dr. Jacobson[31] has set fifty thousand gypsy-moth traps in New England and thus managed to stop this pest imported from Europe from spreading farther over the north-eastern United States. If he has so far not been completely successful in his task of destruction, it is because minor pollutions by not entirely perfect scent molecules have blocked the attractant effect of the perfect ones. Unfortunately, at least with the gypsy-moth, the synthetic decoy substances do have to be of a purity which is technically difficult to attain. This is a fact that has to be faced.

It would be hopeless to try to analyse chemically the odorous substance of every single species of pest in exactly the same way as with the silkworm moth and the gypsy-moth. From the similarity of the attractants so far known, however, scientists have had the idea of starting from another direction as well: they rig up similar types of molecules and see if the males of any species of insect will react to them by fluttering along.

The attempt met with unexpectedly great success right from the outset. In this ingeniously simple way Dr. Jacobson discovered the attractant of the Mediterranean fruit-fly. This pest, which had been imported into Florida, had already turned into a danger threatening the very existence of the peninsula's fruit orchards. Within two years the fly was completely eradicated by means of the combination of attractant and poisonous substances.

After transposition of a few atoms in the attractant molecule, the Oriental fruit-fly responded to the scent. It was just about to devastate the plantations on Rota, one of the Mariana

Islands in the Pacific. Today there is not a single fly of the kind alive there. On Hawaii, a great war against fruit-flies was started, but has admittedly got out of hand. After the destruction of the predominant species, the *Lebensraum* it left free was invaded and conquered by the explosive expansion of the populations of other, until then unimportant species of fruit-flies. It is at present assumed that the sex attractants of about a hundred species will have to be discovered before all the pests can be destroyed.

Shock and Warning Substances

Minnows like to eat their own children; they lack the inhibition most animal parents are endowed with. Instead, nature has equipped these fish with another odd way of behaviour which stops the species from eradicating itself.

Professor von Frisch[32] observed the following phenomenon. Somewhere in the Wolfgangsee (Austria) a school of minnow fry were swimming around, when a large minnow suddenly turned up. The young minnows did not flee, for they are not afraid of other minnows. The grown minnow was overcome by hunger, however, and snapped up a little fish. Now something remarkable happened: all the small fry, having only just escaped, stayed calmly where they were; but the cannibal at once showed panic fear and made off hastily.

Was the criminal seized with remorse for his crime? Such an anthropomorphic idea is quite untenable, of course. Yet in many cases nature has programmed instinctive ways of reaction into the creatures' range of behaviour which may be regarded as very similar to human conduct.

What had happened was this. The moment the big minnow tore up the little one with its teeth, a shock substance was released from the victim's skin and mixed with the water. As soon as a fish of the same species smells this shock substance, it feels an irresistible urge to flee. Only the young minnows remained unimpressed, because sensitivity to the shock substance does not ripen in them until the age of between four and eight weeks.

The curiosity that a predator should put himself to flight

is only a fringe effect of a phenomenon widespread among fish, and one that was really "invented" for a rather different purpose. According to research by Dr. Wolfgant Pfeiffer[33] the shock substance is produced in special club cells within the skin by many species of peaceful fish that swim in schools. The substance is above all designed to scare all members of a school away from the danger area when one of them has been eaten by a predator.

For the loss of a fish is often not registered at all by the other members of the school. The predator lies in wait motionless, hidden among stones and water plants, and takes only a quick snap at a victim swimming immediately past its mouth. It would gradually devour all the members of the school if it were not thwarted by the shock substance.

Dr. Erwin Kulzer[34] describes an interesting experiment with a school of minnows. One fish is caught, its skin grazed slightly, and it is put in a jam jar filled with water. The water (without the injured fish) is then poured back into the lake. What happens among the rest of the school?

> The fish immediately start snapping vigorously and taking big gulps of breathing water through their mouth and gill slits, thereby thoroughly irrigating their nasal grooves as well. They inhale for a short time and suddenly, as if lightning had struck right into the school, they get a shock. A wild dispersal begins, and after swimming this way and that for some time, they either disappear in a hide-out or stay closely crowded together on the lake bottom. After some minutes individual fish shoot up again from the hide-out and chase about in all directions. For days they reject all food.

Every species of fish employs specific tactics appropriate to it. Tenches and crucian carps, if frightened in the aquarium by skins of their own species reduced to pulp, will make violent movements of their fins to crash-dive against the bottom; in open waters this behaviour would make them disappear behind an all-obscuring screen of mud. Gudgeon and loaches become rigid on the spot and will not budge for several minutes; after that their camouflage colour will make them nearly invisible. Striped barbel, which live immediately below the water level, congregate on the surface into a dense school, dart about, and jerk themselves out of the water in an attempt to escape their enemy.

It is remarkable that the shock substance is released only if the skin is injured. A frightened creature cannot actively release it in the way a bird issues a warning call. In other words: in practice, the fish that raises the alarm must die first to draw the attention of the other members of his school to a danger. So the shock substance is not the slightest use to that fish (unless it is only injured), but only to the other members of the species: an extremely altruistic arrangement for the preservation of the species!

Oddly enough, in the realm of fishes nature invented the shock substance once only. Seventy million years ago, in the Eocene, it must have developed in a fish species which was the common ancestor of all carp fish, salmon, and catfish living today. This has been deduced from the fact that it is found only in a single order of fish, the *ostariophysi*, which contains five thousand species, including two thirds of all freshwater fish.

Fish belonging to a different order, which would also benefit from the shock substance, do not possess it: trout, grayling, whitefish, stickleback, pike, eels. Nor do salt-water fish like herring, sardines, and mackerel have any alarm signals by odorous substances. This shows that we should not underrate the role of chance in the inventions that have occurred in the course of evolution.

On the other hand, this invention of nature's has succeeded also with very different kinds of animals. Toad tadpoles, for instance, react very strongly to the shock substance of their own species. With South American water snails too (*Helisoma nigricans*), directly one snail has been torn to pieces by the jaws of a water turtle, the rest start a mass flight by digging into the mud or crawling out of the water.

The toad tadpole's shock substance, however, has no effect at all on carp and other fish that can be scared off by scents. On the other hand, crucians can be put to flight very quickly, and salmon slowly, by the carp's shock substance, and vice versa. The shock substances, therefore, are far less exclusively specific than the sexual attractants of insects. The effect on other species is weaker, the more remote the relationship. This rule works so precisely that on the strength of it Dr. F. Schutz[35] succeeded in establishing some previously undetermined relationships among the five thousand species of this order of fishes. An explanation of the rule is still lacking, though, because nothing is yet known about the chemical structure of the shock substances.

Salmon on their way upstream are warned, though not scared away, by anything that smells of man, bear, dog, seal, or sea-lion. If any of these salmon hunters is waiting somewhere in the shallows, the approaching fish registers its scent as a warning substance and withdraws a little to the side, trying the ascent in some other place.

In contrast to the shock substance, then, the warning substance is given off by the enemy. It is the counterpart in water to a man's or a lion's scent floating through the air to be picked up by deer or gazelle, just as, conversely, a gazelle's scent is picked up by a lion.

Many Africans take it for a sign of intelligence in lions that, like human hunters, they always stalk their prey against the wind. But there is a much simpler explanation: these predator cats, working their way through the high grass of the savannah, take their bearings mainly with their noses so as not to be discovered; whichever smells the other can be sure the other will not smell *him*.

Territory-marking Substances

When an expedition up the Amazon finds a skull stuck on a post in the jungle, all its members realize that this is a boundary stone for the domain of an Indian warrior tribe, beyond which "trespassers will be prosecuted" with fierce hostility.

Rousseau, apostle of "back to nature," wrongly thought that man was turning away from nature by enclosing property. It is a commonplace now that territory marked off as "privately owned" is a frequent phenomenon in the animal world, and that here too there are severe penalties if boundary lines are disregarded: the animals mark off their territory by scent.

Let us take roebuck. During the winter, for expediency's

sake, they join up with many other members of their species to form herds; but by the first half of March they become intolerant of each other. The older bucks dominate the younger, till everyone goes his own way and the winter community is completely disbanded. It is a widely held but mistaken idea that deer live in herds in the summer as well. Apart from the mating season, they are extremely solitary then, and each of these graceful symbols of woodland peace is the fierce enemy of the next.

So in March every roebuck must try to take possession of as large and as favourable a wooded area as possible. He makes the necessary investigations while still roaming round with the herd; attaching particular value to the existence of a dense thicket or a plantation of young firs, where he can find shelter and rest without being disturbed. It is equally important that areas with ample grazing – that is, grassy clearings or fields of grain – should be as near as possible within reach of this thicket.

The conquest of the best woodland involves a number of strangely abstract battles. For the bucks' pointed antlers, their main weapon, are in March not yet ready to fight with: they are still liable to fraying, having skin with a strong blood circulation, which causes the buck excruciating pain at the slightest touch. Consequently, the authority each acquired for himself in battle while with the winter herd, is now respected without a fight.

As the Hamburg zoologist Rolf Hennig[36, 37] explains, the deer of a particular part of the forest know each other quite well, and also know the rank every individual is entitled to. If, for instance, a low-ranking buck investigates a piece of woodland which a high-ranking one has already marked down for himself, the latter need only lay back his ears and turn his head to glance at his rival; with this gesture alone he immediately puts him to flight.

So at a mere turn of the head lower-ranking bucks are driven out of the more desirable woodland, after which they have to hurry to get any territory at all. Directly a buck feels he can stay in a particular area, he at once starts marking the territory claimed by him with a boundary line.

Although this boundary cannot be seen, it is distinctly

smelt on bushes and trees. The roebuck has a scent gland in front of and between the antlers, under the curly hair of his forelock. If he runs this gland cautiously against branches, a strong scent will cling to them for hours.

Since the scent gradually evaporates, the boundary has to be inspected and re-marked up to three times a day. To make things easier and quicker for himself, the roebuck makes the boundary along quiet paths, glades and clearings, as far as he can. Then he need not put down a boundary stone every five yards, as he has to in dense wood, but only about every thirty yards. Thanks to this simplification, he can do a distance of a hundred yards in only a minute or two.

The more roebuck have to live in a piece of woodland and the greater the population density, the more often every individual buck meets his rivals, the more often and carefully he has to mark the boundary line, and so the smaller becomes the territory he can defend. If a buck dies or is shot, immediately the boundary line has evaporated, his neighbours extend their territory into the deserted one until they meet each other, and then proceed to draw new boundaries. So the size of a roebuck's territory varies between 25 and 250 acres, depending on population pressure, ranking and the character of the terrain.

If man's nose were better, he could scent considerably more territorial boundaries in the forest. The female deer, the doe, is also solitary. The frontiers of their territories overlap with those of the bucks, which does not imply any troth for the mating season later on. Absurd as it sounds, despite the intensive protection of does by foresters and gamekeepers, further details about their behaviour in their territories and the marking of boundaries are still unknown today.

Other scent markers are the hamsters,[38] which indicate their boundaries by the scent of their flank glands; martens and badgers, which use the secretions of their anal glands for the same purpose, and voles, which squirt their secretions onto their soles to imprint scent marks along the boundary by a drumming movement of their legs. Brown bears have hitherto been suspected of suffering from mange because they keep rubbing their backs and muzzles against trees and rock edges. Now we know that this is no disease but the marking off of territory at prominent points in the landscape. By rubbing, the bears leave a strongly scented, greasy trail. To ignore these marks spells mortal danger to all other bears.

Wild rabbits were also thought to have mange, although people should have noticed that the place where these animals scratch with their hind legs is nearly always their chin. After closer observation Australian naturalists[39] found out that there are neither mites nor fleas on a rabbit's chin, but scent glands, the secretions of which are also stamped on the ground by their feet when the rabbits inspect their boundary lines. When the population density is extremely high, the males rub even their females and young with the odorous substance in order to make them known unmistakably as their personal property to all their rabbit neighbours.

Another source of scent is used by many other animals for the same purpose, namely droppings and urine. Even a town-dweller gets some idea of this when he takes his dog for a walk. All the tree trunks, lamp posts, and corners of houses are sprinkled with minute portions, not because the dog has to keep "relieving" himself so often, but because at these points he is fighting out ritualized scent duels with mostly invisible rivals for a no longer realizable ownership of territory.

How lucky that we do not have to take hippos for a walk. Professor Bernhard Grzimek[40] writes that

> they, too, have their fixed territories. The frontiers are fairly rigid and must not be crossed, because the adjoining land belongs to other animals of the same species. So they are not much better off than we men with our customs frontiers and passports. Hippos mark their territory by scent marks in that the bulls fling about their dung for yards, using their short tail like a propeller and adding their urine, which they squirt diagonally backwards. Thus small bushes or clearings are sprinkled with the dung, which does not have too unpleasant a smell even for humans, and any strange hippo that might contemplate immigration knows immediately this is occupied territory. The intruder would have to fight terrible battles if he wanted to stay here.

The spotted hyena drops little lumps of faeces round its prey in case it has to interrupt its meal; so as to indicate to all hyenas arriving later that it has found, and therefore owns, the carcass. Foxes, too,[41] mark their territory in the woods

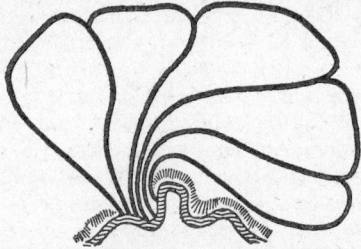

Figure 53. Patterns of territorial scent marks of some hippos on the bank of a river in the Congo. Since the steep part of the bank is impassable for the bulky animals, the neighbours allow each other the use of a "corridor" giving access to the water.

by urine. In the same way they perfume their females, so as to prevent any doubts on the part of the neighbouring foxes. Tree shrews first prepare a little puddle of urine, trip about in it, then march along their boundary for as far as their "stamping ink" lasts. Then they start again from the beginning. Many lemurs first urinate into their forepaws, rub the liquid on to the soles of their feet, then start patrolling and marking their frontiers.

With domestic mice,[42] the perfuming activities sometimes even produce distinctly visible marking stones. These rodents, too, define the territory of each family in warehouses, larders, and cellars by markings. In every spot where neighbouring, that is, warring families meet, a boundary stone is laid in the shape of a lump of faeces. If the mice are not disturbed by rats, cats, or men, constant urination turns this into an oddly shaped, padded pillar looking like a stalagmite in a cave and reaching a height of up to two inches.

There is also the very interesting possibility, proved by P. Leyhausen and R. Wolff, that the distribution of animals of the same species over the available living space can be ruled not only by rigid frontiers, but also by an elastic time-

table. According to Professor Konrad Lorenz[43] the two
scientists have

> found that in domestic cats living free in open country,
> several individuals could make use of the same hunting
> ground without ever coming into conflict, by using it
> according to a definite time-table, in the same way as our
> Seewiesen housewives use our communal wash-house.
>
> An additional safeguard against undesirable meetings
> consists of the scent marks which cats will deposit at
> regular intervals wherever they may be. These have
> exactly the same effect as the block signal on the rail-
> ways, designed to stop two trains from colliding. The cat
> going hunting that comes on the signal of another cat,
> the age of which it can well judge, will hesitate or change
> its route if it is a fresh deposit; or calmly pursue its
> course if the deposit is a few hours' old.

Messenger Substances

It is only since about 1958 that animal psychologists and
biochemists have started a detailed investigation of a strange
phenomenon. Insects living in social communities like ants,
termites, bees, wasps, and bumble-bees secrete hormone-like
odours in order to exchange messages and orders among
themselves. As soon as these are smelt, the substances either
immediately produce the behaviour demanded by the insect
exuding the scent in the subjects of its state, or they even
fatefully change their bodily structure for their entire lives.

So these individual odours take their effect not like regular
hormones in the inside of an individual that may produce
them, but outside. They join two or more individuals into a
higher unit. Eugen Marais spoke of the "soul of the white
ants," which really only made things still more mysterious; and
other scientists too have shown a tendency to compare the
insect states to a "super organism" consisting of thousands
of millions of insects. Messenger substances do play a role
similar to that of the hormones inside our bodies, and they
should help to solve the great riddle of this higher unity in
insect states.

All hormones with an outside effect on other members of the species were at first called ectohormones, but since 1959 the biochemist Professor Peter Karlson[44] of Marburg and the entomologist Martin Lüscher[45] of Berne have referred to them by the more specific name of pheromones. They include the sexual attractants discussed above (pp. 130–8). These scents, however, can achieve a surprising amount more than the mere attracting of sexual partners.

The American zoologist Professor Edward O. Wilson[46] of Harvard University found indications of a regular scent language in social insects. He has already deciphered the first words of an apparently copious dictionary of odorous substances, and he even suspects that ants might have a kind of syntax. Considering all he has found out so far, such a possibility seems perfectly feasible.

Already, at the very start of this new branch of research, it emerges that the animals' scent language is not just a primitive substitute, conditioned by the medium, for the language of sounds and gestures, but an amazingly versatile, ingenious, and highly differentiated means of communication. With ants alone ten different scent "words" have been deciphered, opening up new insights into the social organization of these insect states.

When an American fire ant finds, some distance away from its nest, a dead butterfly which is too heavy for the ant to carry off alone, it will quickly run homewards to fetch help, marking the way back by a scent, rather in Boy Scout style. The ant pushes its sting out of its abdomen and presses on the scent gland, making a fine trickle of the liquid run along the

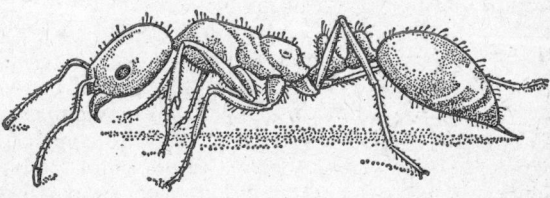

Figure 54. A fire-ant worker 'writes' a scent track on the ground with its sting.

sting like ink along the nib of a fountain pen. But the ink does not write a continuous line on the ground. For one thing it would be wasteful; and for another, ants coming across the trail would be unable to recognize which of the two directions they are meant to follow. This is why the successful forager draws a shaded line. Every part of the shading is shaped like an arrow pointing in the direction of the prey. Unlike a human signpost, of course, the arrow point is not drawn consciously, but is formed by gradually increased pressure and a sudden upward jerk of the sting. But it is all the more remarkable that the arrow, which cannot be seen, only smelt, is interpreted correctly by the other ants. As there are innate optical patterns of recognition (see p. 55), there are apparently also olfactory ones.

The scent rising into the air from the hair-thin lines is capable of attracting other ants from both sides at a maximum distance of four-fifths of an inch. The ants at first make directly for the trail and then march in single file exactly along the markings up to their objective. A red ant, however, crossing the trail of a fire ant, will pass it blindly. Each of the many ant species has its own secret trail scent, which cannot be registered by the members of strange species.

The trail scent of a fire ant evaporates within two minutes. During this time the insect can crawl a distance of sixteen inches at the most. In a way, therefore, it drags along a scent "tail" that can never grow longer than sixteen inches. At first sight this seems like a huge drawback to the technique of scent substances, but it is not really one. For the homing ant is trying to make its trail as straight as possible. The pathfinders which follow her appreciate this and do not let themselves be confused when they reach the evaporating end. They keep on running straight ahead for a bit and so also reach their objective unless it is impossibly far. Besides, the trail does not *have* to run right into the next; it only needs to reach a route much frequented by ants or a neighbourhood where many fellow inhabitants of the same ant-hill are to be found. It would be a disadvantage, in fact, if the ant trail were of a more lasting kind. Suppose the route leading to a single dead butterfly could be scented for an hour; an army of thousands of ants would head for the spot continuously all that time, and only the first arrivals would find anything left.

Nature has here achieved a miracle of economic rationalization with a minimum consumption of scent substances: the farther away the source of food is from the next, and the smaller its size, the fewer workers are informed about it. In each case there are just as many of them as is necessary and desirable.

If the source of food is worthwhile, paths will very quickly appear, with markings constantly renewed by countless ants, and scented intensively enough to cause a mass influx. But as soon as the source of food is exhausted, the scent of the trail disappears too, since ants which go empty do not leave any more scent arrows behind them. So the stream of hungry insects dries up at once.

An ant colony needs on average for its daily consumption about four and half pounds of insects – a considerable amount. To find, hunt and collect them without using up more energy than is gained from their consumption, the most finely balanced tactics are necessary. Dr. D. Botech[48] once put all contributing factors into mathematical correlation, making a numerical check on nature's cunning. The result of his calculations was that the ants work with such optimal efficiency that even a computer could not devise a better procedure for them.

The dreaded African driver and South American army ants, whose nomad armies roam the land, feeding on everything which does not flee in time, from the aphid to the python, apparently have a complicated network of scent messages. It is an established fact that the reconnoitring vanguard leaves a trail scent; and presumably the main army receive scent messages from which they learn whether they are to wait, advance, surround a victim, or carry out other operations.

In the earth hole of the leaf-cutting ants, scent trails lead to ingeniously cultivated underground mushroom gardens, the main source of food for these insects. Because of constant use the scented "threads of Ariadne" are installations lasting for several months. They show every hungry citizen of the ant colony the way to the dining-room. Scent markings in the interior of the mazes with their hundreds of tunnels generally seem to be the signposts showing the insects which is the way to the queen, the breeding colonies, and the outside world. But in this field much research is still needed.

Scent substances are also used in an ant colony to raise the alarm. To man these pheromones smell mild and pleasant; but they are capable of rousing large colonies of ants to immediate furious action. However, if only a single ant is attacked – for example, by a predatory beetle – it would not be expedient to stir up the whole colony because of it. So the alarm system, as Professor Wilson has established, works in the following way:

The excited insect sends out a special alarm scent. Some species of ants simply secrete it from a gland and leave it to evaporate on the surface of their bodies. The central European black ant lowers its head and points its abdomen steeply upwards. The droplet of scent substance emerges from the highest point and spreads to take effect. Other ants (and wasps too) curve their abdomens forward underneath their heads and squirt their alarm substance precisely at the enemy, which in this way is branded.

The alarm scent spreads evenly in all directions. Between a distance of one and half and two inches from the alarm-giver (or from the branded one) the outer scent layer, since the pheromone is fairly well diluted, acts only as a signal attracting attention and makes all ants within its reach run towards the scent source. When the insects have come within a distance of one inch, but not till then, they plunge into a highly concentrated central region which releases in them the frenzy of alarm, makes them send out their own alarm substances, and renders them highly aggressive.

So minor local disturbances can very well be cleared up locally as well. Meanwhile the speed of the running ants, the spreading velocity of the scent spheres, and their evaporation are so well attuned to each other and to the creatures' threshold of sensation that – with a certain safety margin – the fighting troops are always mobilized at about the strength needed to overcome the danger.

Although the alarm substance causes the ants inside the nest to attack, outside the nest it makes every ant that smells it take to flight. Other pheromones stimulate the ants to go out foraging, increase the size of their nests, feed the queen, look after the brood, clean the body of a comrade, or give food to another that is begging.

Even dead ants still produce a very important pheromone. A worker that has just died (it may reach an age of up to ten years) is cared for by other workers as if still alive. Its immobility and twisted shape do not make the slightest difference,

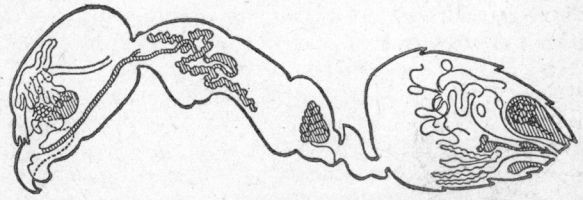

Figure 55. The "vocabulary" of an ant is contained in several scent glands (shaded), which produce various kinds of pheromone. Each scent has a particular signalling value. Professor Wilson thinks it possible that a combination of two or more scents may be used to convey yet another meaning.

for during the winter rigidity this happens to all ants. But after a day or two special pheromones are formed in the decaying process, which in the living creatures produce the funeral reaction. They carry the carcass to a refuse heap outside the nest.

When Professor Wilson perfumed live worker ants with this pheromone, they too were carried to the cemetery, although they put up a valiant resistance. There they rallied, hurried back to the nest, and were immediately carried off again. This to-ing and fro-ing went on about thirty times, until the smell of death had worn off their bodies.

Professor Wilson has also found indications that ants can combine various scent substances into a mixture, thereby acquiring a larger vocabulary of signals than would be involved by the number of scent glands in their body. Probably they can also send out pheromones in varied impulse sequences, and modify them by changing their intensity, thus creating a sort of code. In such circumstances the existence of some kind of syntax is at least conceivable.

Another possibility is a combination of scent signals with sounds. Many ants produce squeaking or gnashing sounds with their abdomens by rubbing a segment of their bodies against special grooves on their bellies, as if against a wash-board. They also make a cracking noise with their mandibles and knock their heads rhymically against stones. The California termite *Zootermopsis angusticollis*[49] hits its head against the ceiling of its nest galleries. In case of disturbances or attacks from outside, it induces its fellows by this warning signal to withdraw into the deeper parts of the structure. To avoid confusion with any chance knocking noises, the termite uses a code in which sequences of two or three single raps follow each other with a short interval in between.

So far we do not know anything about the significance of ant and termite sounds. But many mysterious phenomena in their social organization, many bewildering capabilities and behaviour patterns of the members of insect colonies, will presumably find surprising explanations as the deciphering of their vocabulary proceeds.

Scent Substances That Change the Body Structure

A state threatened by enemies may meet its doom in two ways. If it has not enough soldiers, it will succumb to the enemy's attacks. But if it has too many soldiers, its economy will break down under the mass of unproductive members.

Termite states, too, have to defend themselves with soldiers against predatory ants. It is vitally important for them as well that their army, which cannot feed itself, should be neither too small nor too large. How do the insects set about adapting their recruitment exactly to their economic con-ditions? Nature has devised a regulating mechanism similar in effect to that produced by human thought.

Basically, all termite workers are potential soldiers.[50, 51] Without measures to preserve this social equilibrium, there would be only the royal couple, reproductives, and soldiers, all of them doomed to starvation. That is why the soldiers constantly secrete a special scent substance which has a profound effect on the development of the rising generation, the so-called nymphs. As soon as the scent concentration exceeds a certain threshold value, the nymphs' growth glands atrophy, and instead of maturing into fearful giants with huge and frightening pincers, they remain small and modest, just workers. But if the termite army has suffered severe losses in a battle with ants, the lack of soldier scent very soon turns many nymph dwarfs into giants – just as many, that is, as are needed to fill the gaps in the army.

There is thus a kind of continuous census going on the termite state. Its results are shown in the concentration of scent substances within their fortress, which can indicate one of two things: there are either too many soldiers or too few. Depending which it is, a smaller or a larger number of new soldiers grows up. The size of the army is thus constantly fluctuating round an economically optimal level. In the course of evolutionary history nature has worked out what this level is and set the controls accordingly by harmonizing together the scent secretions of the individual insects, the dilution of the scent in the air of the termite fortress, and the sensitivity of the growth glands to the scent's effect.

With other castes too the employment office of the termite state works in a similar way.

First of all it was believed that the scent substance would have to be eaten by the termites together with the food, would then take its effect like a hormone, and would by-pass the nervous system to control directly the development of the growth glands and thus the growing insect's size, shape, and purpose in life. In 1961, however, J. Pain,[52] a French woman scientist, discovered something quite different, at least among bees.

The queen bee represents, as it were, a one-person caste within her state. She therefore prevents any rivals from growing up by secreting the queen substance from her mandibular gland. This scent inhibits the development of ovaries in all her worker bees and prevents ovary development in ants, flies, and termites, and even kills mosquitoes.[53] It works, too, as a means of controlling pests in the beehive.

But queen substance is much more advantageous than any

man-made insecticides, for it is quite enough to suppress ovary development if the bee workers merely smell the scent with their antennae. Everything else follows automatically by way of the activities of the nervous system. Just imagine what it would be like if by constantly smelling a perfume, people could remain immature, like children, all their lives.

A pheromone of the migratory locust also works as a magic perfume, altering the structure of the body. Normally, the members of these insect species capable of devastating countries are harmless small grasshoppers, hopping about here and there without giving any hint of sinister purpose.

But suddenly the mass invasion takes place. Professor C. B. Williams[54] of the British experimental station at Rothamsted observed a swarm consisting of some ten billion dun-coloured creatures in East Africa. The wing span of the locusts was six inches. The immense swarm, over one hundred feet deep, eclipsed the sun along a front of a mile and a quarter. Even if Professor Williams had managed to destroy a million locusts a minute, it would have taken him seven days and seven nights to master the invasion. He had to look on helplessly while the land was eaten as bare as if it had been ravaged by a fire.

How are these hardly noticeable grasshoppers turned into such devastating swarms? By a scent of togetherness. But as this pheromone is very volatile in the open, it works only over a distance of an inch or two from insect to insect. Nothing happens until a critical density of population has been reached; not all of the conditions for that have been certainly established. First climate and the available amount of food must favour a particular increase in the locusts' numbers. Then they are often swept away by winds. In tropical and subtropical countries there are regions where north and south winds clash; and here the locusts, which are still harmless, are blown together like paper in the whirl of a street corner.

Then they seem to have a tacit agreement that trees and other outstanding landmarks are their meeting points. Thus a congregation gradually gathers in these places. The critical distance of the pheromone effect is passed, and what happens now is the exact reverse of the self-limitation in the termite state: the more locusts get together, the more fertile they become. The first generation produced here represents a transition form. A reproduction avalanche, a population explosion of unimaginable size takes place in so short a time that as a rule in desert areas no one notices anything about it. The second generation then grows into large, long-winged specimens, which no longer have any resemblance to their grandparents. They become migratory.

Before they can fly, the locust larvae form into bands. The masses are in the grip of a nervous desire for action. Separate bands join up to form an army. The hopping of small groups spreads like a wave over the whole community and makes others follow suit. The even rhythm of movement thus created eventually results, as Professor Adolf Remane[55] puts it, in a forced march in a fixed direction.

What happens then was observed by the French zoologist Rémy Chauvin[56] on the island of Corsica. The youth army of migratory locusts in their billions, still unable to fly, carried on their voracious march across country, sticking to the same direction day after day. Coming to insurmountably steep rocks, the ones marching in front stayed jammed together, while the rest of the columns pressed on irresistibly, making a locust ramp from which all the others could negotiate the obstacle. In the same way living bridges were formed across minor waters. Fires lit by people to stop the invasion were smothered by the charred bodies of the vanguard, producing a breach through which all followers could advance farther. These insignificant insects had been turned into a vast and tremendous force by nothing more than a delicate scent.

Whereas here the scent triggers off a population explosion, in other creatures social scent substances are capable of preventing overpopulation.

If a large tadpole is put in an aquarium with a group of small ones, they are overcome by a strange and fatal loss of appetite.[57] Though offered plenty of food, they suddenly stop feeding, and soon die. The same thing also happens without any big tadpoles being there, if the water some big ones have been swimming in is poured into the little ones' tank. So this must be due to a secretion whereby nature arranges for a sort of privilege for the first-born.

As with the termites' soldier scent, the secretion of the inhibitory substance, the dilution by water, and the sensitivity of the tadpoles are precisely attuned to each other, and in such a way that there are never more frogs growing up in a pond than can find sustenance there under normal conditions. The population explosion is "defused" by the scent substance. In this case one might almost speak of a physiological providence, for it is not today's hunger but the threat of tomorrow's that triggers off the regulation.

This is something extraordinary. Death from hunger, beasts of prey, the elements, and disease used to be considered the only regulating factors for maintaining nature's balance. But these are only of secondary importance, in some cases even incidental or superfluous. Curiously enough, animals' self-limitation of their numbers was not recognized as the decisive factor in preventing overpopulation until 1962, which was when the contraceptive pill first came into prominence along with other measures of human population control.

Besides contraceptive scents and the effects of stress and degeneration in social behaviour, nature's measures comprise everything from simple abstention, by way of a *numerus clausus* excluding supernumarary members from the business of reproduction, the parcelling out of living space, and the inhibition of growth, right up to cannibalism. The Scottish zoologist V. C. Wynne-Edwards,[58] Professor at Aberdeen University, should have the credit of first recognizing all this as a widespread principle of nature.

In this context I can deal with the subject only as far as scent substances play a part in animals' controlling their own population density. The flour beetle is a very impressive case in point. These insects, a pest dreaded in the flour stores of the whole world, propagate rapidly. As soon as the population density rises above two beetles to a gram of flour, the females devour their eggs immediately they have laid them. The beetles' excrement contains a chemical scent substance, which has a cumulative effect in reducing the fertility of the entire population in a storehouse. It also prolongs the time in which the larva, the well-known mealworm, develops into a beetle, and as the concentration grows even stronger, it triggers off egg-cannibalism.

Sensational experiments with mouse pheromones have been carried out by the Dutch scientists S. van der Lee and L. M. Boot since 1955. The more female mice they put in an enclosure, the more sterile these become. Cases of pseudo-pregnancy increase till the menstruation cycle of the animals is completely blocked. This fertility inhibition can be eliminated in three ways: operative removal of the scent glands; transferring the mice into individual cages at a good distance from each other; or mixing male mice with the females.

There are some complications, however, about the scent of the male mice. Dr. Helen Bruce[59] at the London National Institute for Medical Research found that the male scent removes the female's fertility inhibition only if it is her partner's. If a pregnant female is joined in her cage by a strange male, his scent interrupts the development of the embryo which has already started. Unfaithfulness on the parents' part is fatal for mouse embryos. But unfaithfulness is also a sign of degeneration, which always appears in times of overpopulation.

The Artificial Nose

None of the experimenters knew who the subject of the experiment was, except that he had been there before. He was carried into the laboratory on a litter, masked and hooded like a member of the Ku Klux Klan, and was pushed into a horizontal glass bottle over six feet long. After the lid had been screwed on firmly, there was a humming noise. A ventilator fanned air at him which had been made completely free from smell, while at the other end of the bottle the air contaminated by the body's secretions was sucked off and passed on to an "artificial nose," a gas chromatograph.

After about half an hour the apparatus had divided the human smell into twenty-four different fractions and defined the quantitative contribution of each of these substances. This scent register can now be used as a personal identifying mark, just like a finger-print or a photograph. The experimenters looked at their index, comparing the scent register with their records of previous scent analyses of people, and very soon reached the conclusion that the man the machine had been sniffing at could only be Mr. Peter Morgan.

The inventor of the apparatus, Dr. Andrew Dravniek,[60] had high hopes after the first experiments in the autumn of 1965. He was confident that an improved version of the machine would soon be turned out by the laboratory of the Illinois Institute of Technology at Chicago. The police would then have a means at their disposal, on the strength of the scent visiting card which the burglar leaves behind in a closed room, to find out the culprit among a number of suspects. Since people related by blood have similar body smells, the proof of paternity might also be carried out by the sniffing machine.

The use of the machine in medicine would have incalculable advantages, Dr. Dravniek believed, for if each disease is accompanied by characteristic body secretions, a scent analysis of the patient would provide an early and reliable diagnosis. In another field, a submarine hunter could be set on the backwater scent in the wake of an enemy submarine, so that the hunter would follow it as the male sperm whale follows the scent of the female.

Clearly, the artificial improvement of human smelling capacities would be of revolutionary importance, with all sorts of other practical applications easily imaginable. But it is also clear that human technology is still ridiculously clumsy in meeting the problems involved in developing such an artificial nose. A dog's nose or a mosquito's antenna can do in a second what has been achieved in half an hour by the sniffing machine, an expensive apparatus on which enormous time and effort has been spent.

The way the machine works is quite unlike our process of smelling. Other "electronic tracking dogs", which have so far been used, also work on "unnatural" principles. The automatic gas-detector, which is designed to give warning of dangerous gases, vapours, and smoke development in fuel depots, hospitals, mines, and ships' holds, sucks the air between an ultra-violet lamp and a photo-cell. As soon as this registers a change in the strength of the light received, it sounds the alarm.

Nature uses other and better ways, only we do not yet know which. Professor Dietrich Schneider,[61] Director of the Max-Planck-Institute for Behavioural Physiology at Seewiesen, characterizes the present situation thus: "We know only the stimulating substances on the one hand and the sensory perceptions on the other; we are still in the dark about the intermediate process of stimulating the sensory cells by odorous substances."

There are dozens of theories about this, but so far none of them has been made to square with the facts – not even the hypothesis of the three Americans, John E. Amoore, James W. Johnston, and Martin Rubin, [62] which attracted a lot of attention in 1964.

The three scientists suggested that the stimulation of the olfactory sensory cells by scent was not a chemical process at all, but rather had something to do with the size and geometrical shape of the scent molecules. According to these scientists, in the receptor there are tiny holes of various shapes, smaller than can be resolved by the electron microscope, and a corresponding type of molecule fits into every one of these like a key into a lock. They say that we experience:

spherical molecules as camphor-like;
disc-shaped molecules as musky;
disc-shaped molecules with a "tail" as flowery;
wedge-shaped molecules as like peppermint;
rod-shaped molecules like ether;

and that many other scents are composed of these elementary smells. Just as any shade of colour can be mixed from three basic colours, many qualities of smell can be composed from a few basic smells. But we need only think of substances with a pungent or decaying smell to see that not everything can be accommodated within the scheme.

Its inadequacy becomes even more evident if the process of the olfactory sensation is more closely investigated. Professor Dietrich Schneider[63,64] carried out this task on the antenna of a butterfly "because it is much simpler to survey everything from here than in the mucous membranes of dogs' noses."

This is what "simpler to survey" means: the stem of a

silkworm moth's antenna, which is only a quarter of a millimetre thick, contains no less than forty thousand fibres.
Thirty-five thousand of these conduct signals from olfactory
sensory cells to the brain, while five thousand pass on different
sensory messages. What is required now is to record from a
single fibre of an olfactory nerve with a microelectrode in
order to listen in to the messages in it, while the antenna is
stimulated by different scent substances.

*Figure 56. So small is the world in which our investigators of the
senses work. The section of the butterfly's antenna framed on the
left is shown enlarged in the middle with a rectangular section, of
which a further enlargement is shown on the right. Some of the
sensory cells which become visible only in the last enlargement must
be tapped for listening in to their messages.*

What transpires at first is a surprising similarity to the signal
code of the visual cells (see p. 23). It is possible for three
adjoining cells to react to one and the same scent substance
in entirely different ways. Cell A steps up the volley sequence
of current impulses, cell C reduces them, and cell B does not
react at all.

When Professor Schneider had examined a few hundred
olfactory cells, he found out that among them there are

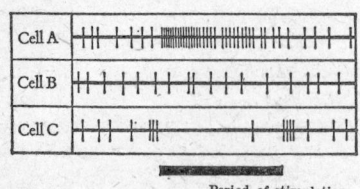

Figure 57.

extreme "specialists" as well as "generalists." These names
already point to the fact that we are still a long way from
exact knowledge. "Even those specialists which we can under-

stand more clearly cause difficulties," says Professor Schneider. "Technical expenditure for the differentiation of these functions is steadily growing and occupies our whole team. Before we can make more exact statements, we shall just have to wait for what insights our further work may bring." So I shall here give just a brief outline of the position reached so far.

The specialists react only to a single scent substance or only to a very small group of scents. In the male of the East Asian silkworm moth *Antheraea pernyi* these sensory cells react to the sexual attraction of the female moths, which to us is odourless, by quick sequences of volleys, while some other smells have a retarding effect on the repetition rate.

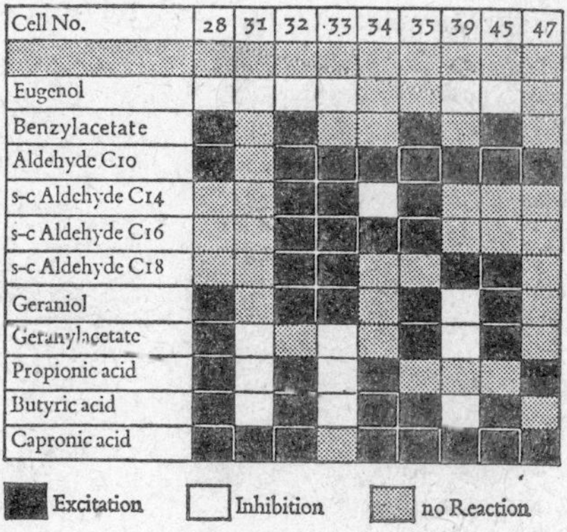

Cell No.	28	31	32	33	34	35	39	45	47
Eugenol									
Benzylacetate									
Aldehyde C10									
s-c Aldehyde C14									
s-c Aldehyde C16									
s-c Aldehyde C18									
Geraniol									
Geranylacetate									
Propionic acid									
Butyric acid									
Capronic acid									

■ Excitation   □ Inhibition   ▨ no Reaction

Figure 58.

Another such specialist was found by Professor Schneider's pupil Dr. Viet Lacher[65] on the antenna of the honey-bee. Some of its olfactory cells react only to carbon dioxide. A change in the carbon dioxide content of the air inside the hive from 0.5 to 1.0 per cent leads to an acceleration of the rate of firing to 80 per second: an extraordinarily sensitive registration device, which, we can only assume, serves to regulate the supply of fresh air by ventilator bees! Incidentally, warm-blooded animals and man have a similar carbon dioxide meter[66] in the arteries of the brain stem. It controls the rate of respiration.

The reactions of the generalists, however, are unexpectedly versatile. Figure 58 shows the results with a number of olfactory cells which, like our noses, are sensitive to a variety of scent stimuli.

Each cell has a different selection of scents it reacts to. The number of scents offered may be increased indefinitely, yet no regular system emerges. But examine as many ce ls as you like; no two with identical scent spectra have so far been found. A butterfly's antenna has about twenty thousand receptors for general smells, but probably each of these has different characteristics in what it answers by excitation, inhibition, or no reaction at all. This is shown by the vertical columns of Figure 58.

From the horizontal lines of Figure 58 we can see which messages are sent by the individual cells to the brain when a particular scent substance is present. If the table is enlarged to comprise thousands of olfactory cells, we should get a rather confused code of simultaneously received excitations and inhibitions. We know for sure that insects recognize these scents very well and that they can therefore understand this code. So at present we are still completely baffled by the problem of how the tiny brain of a butterfly or bee can decipher this mass of signals.

The insects' sensory cells form from the chitinous exterior of their bodies. A long fibre grows from a branch of the antenna to the brain. A short fibre advances in the opposite

Figure 59. If the right side of Figure 56 is enlarged still further, one can see among other things the sensory haircells which serve the silkworm male as receiver for the sexual attractant. Enlargement of a section of the long haircell shows the decisive factor: from ultra-fine pores emerge nerve endings so minute they can be seen only under an electron microscope.

direction, establishing contact with a sensory cilium, a pore, a sensillory pore, or some other type of sensilla. It travels, for instance, together with fibres from two or three other sensory cells, into a sensory cilium, unravels itself into several ends, exposed to the environment with its last tiny tips through some ultra-fine pores of the cilium.

These last ends of the sensory nerve cells are presumably where the mysterious process transforming the physico-chemical stimulus into biological stimulation takes place. Until these riddles are solved, our dreams of an artificial nose will not be fulfilled.

The Sense of Taste

Merely by intuition we are inclined to think that the sense of taste must be something similar to the sense of smell. With the lower animals, particularly with molluscs that live in water, the two senses are in fact the same. They have nothing but a uniform chemical sense. But with all higher animals, including the insects, nature has made a complete division. A crude "near sense" is designed to examine the food by direct contact, while a highly sensitive far sense is receptive to scent even from a great distance.

Before we can experience the faintest taste, twenty-five thousand times as many molecules have to affect our tongue as would be needed to stimulate our nose. But qualitatively, too, our sense of taste with its basic sensations of sweet, salty, sour, and bitter, is miserably poor. All delicacies which at an exquisite dinner are so relished by our palate, are in fact not even registered there, and only very roughly by the tongue. For here again it is the scents rising from the cavity of the mouth into the nose to which we owe our gourmet's delights. To experience a delicious taste, the sense of taste itself is quite secondary.

Moreover, the tongue does not even tell us anything about the chemical nature of the object it touches. One and the same substance may taste totally different, depending on its concentration. The salt potassium bromide, for instance, at a concentration of 0·01 moles tastes sweet, at 0·02 bitter-sweet, at 0·04 bitter-salty, and 0·20 salty. Magnesium sulphate, another salt, which formerly was often used as a laxative, produces a salty sensation at the tip of the tongue, while farther back on the tongue it seems bitter.

Still, we must not be too hard on the sense of taste! Apparently it is due to the circuits of the taste nerves that they report to the brain much less of the processes on the tongue than they actually register. For, by reflex action, they carry out the most subtle control of the composition of the saliva. Depending on the kind of food that enters our mouth, the taste nerves immediately change the production programme of the nearby salivary glands in such a way that the food can best be digested. In many cases the saliva has already changed before we, with our subjective experience of taste, notice any differences in the quality of the food.

So besides the sense of taste we are conscious of, there exists a reflex system more important to us and – fortunately – also more efficient; but it is still largely unexplored.

Chapter Five

The Sense of Hearing

Adaptive Forms of Acoustic Sense Organs

IF man could hear in the ultrasonic range, he would certainly have to stop up his ears with wax at night. Otherwise, at least in the country, he would never be able to sleep for noise. Nor would poets have ever extolled the peace and quiet of the night. Their nerves racked by the constant cries of battle, by the ceaseless clack and clatter, they would have to look elsewhere for silence, perhaps amidst the ordinary traffic of a big town.

The originators of the ultrasonic nocturnal din, so fortunately inaudible to us, are the bats and nightmoths. They conduct unceasing war against each other with weapons, counter-weapons, and counter-counter-weapons, listening in to enemy broadcasts, jamming radio transmissions, and employing other refinements. Here we find sensory organs and behaviour patterns adapted in the most astonishing ways to the enemy's technical inventions. It is such adaptations alone that make survival possible.

Most fruit-bats or flying-foxes have not acquired the faculty of echo-location, although they are creatures of the dusk. Ivan T. Sanderson[1] describes this as follows:

> Fruit-bats often sleep and live together in huge populations. Punctually at dusk they make for their feeding-

places, which may be up to twenty miles distant. On their way there they even cross over the sea if they are bound for an island. In closely crowded swarms, in a stream which sometimes sweeps across the sky for hours and hours, they flutter with slow wing-beats through the clear evening sky, taking no notice of all the human activity below them. To meet such a swarm of fruit-bats, numbering millions, is one of the most impressive events of a journey to the tropics. Wherever the swarm alights, there is a never-ending squabble about the feeding-place. When feeding, the fruit-bat hangs head down, looking somewhat like a ham. It holds the fruit it is eating with one of its hind legs, while the other clings to a branch. Those which have found their place defend it against later arrivals by blows with their strong thumb-claws and by bites. The procedure is the same towards morning when the whole host fly back to their sleeping-trees.

At dawn and dusk the fruit-bats take their bearings by means of their particularly large, light-sensitive eyes. But in total darkness they are completely helpless. The fruit they feed on they have to find on the trees by touch and smell.

By comparison, the tongue-clicking of the tailed fruit-bats of the genus *Rousettus* signifies an immense step forward.[2] The noises produced, which are audible to man as well, are eminently suitable for echo-location. So these fruit bats can also fly about on a dark night and change their feeding place.

Can this simple process not be made use of for blind people? Attempts have been made to equip blind people with castanets – unfortunately without much success. Donald R. Griffin[3, 4] of Rockefeller University explains the failure as follows: In the bat's midbrain there is an auditory centre that still recognizes sonic impulses as separate sounds if there is an interval of the thousandth of a second between them. This means that the creature can register as separate noises sonic impulses which shoot through the air at a distance of only thirteen inches from each other. It can thus distinguish by the echo alone between two objects which are only a little over a foot away from each other. So fine are its powers of producing an acoustic image.

Moreover, a thousandth of a second after a bat has sent out an exploring noise, its ear is already capable of receiving the first echo. This means that even near objects, seven inches at most from the nose, can be located. This is especially important for catching insects.

What is decisive for echo location is therefore not only the presence of a transmitter, whether a clicking tongue or a castanet, but above all that of a suitable receiver. This, unfortunately, is just the weak point in human hearing; or rather, human hearing has been specializing in other things, since echo location is not among the necessities of our life.

Our auditory nerves are much more inert. In particular, we cannot register soft echoes directly after a loud signal has been sounded – as one of Donald R. Griffin's experiments shows.

> You can easily convince yourself of this by playing a tape recording which contains sharp "tick" sounds with the corresponding echo. The recording should have been made in a closed room. If the sound is examined with physical instruments, it can be found that every "tick" sound is followed by several echoes, which come from the different walls. But we cannot hear these. If the same tape is played backwards, however, so that the weaker echoes always occur before the "tick" sound, we can distinctly hear the echoes.

But, compared to the flying foxes, bats have also considerably improved their transmitters. They do not click their tongues but produce in their larynx ultrasonic cries of a very high sound intensity, which they emit through their open mouths. The little brown bat, the largest Central European species, normally transmits in flight about twelve clicks a second. When it approaches its goal, the ultrasonic screams speed up steadily, reaching 300 bursts a second. The speed of machine-gun fire is slow compared to it.

Nor is the comparison exaggerated as far as loudness is concerned. Four inches in front of the bat's mouth Griffin measured 100 decibels. (A pneumatic drill working in the road makes a noise of 90 decibels. The pain threshold of our brain starts at 130 decibels.)

Why do bats cry in tones too high for us to hear? Not just to let human beings sleep in peace! It is because ultrasound can be bundled and beamed like a searchlight, and so yields a good deal sharper echo "images" than ordinary sound. The little brown bat's cry varies from 100,000 to 50,000 vibrations a second, which corresponds to wavelengths from 3 to 6 millimetres. This is what makes it physically possible for even tiny flies and mosquitoes to produce an echo. They give no reflection at all of human cells, which as a rule have wavelengths of between 10 and 5 metres.

No one has yet established why many bats do not irritate each other in the regular Babel of cries from the swarm. A single shelter, for example, the Ney cave in Texas, used to hold over twenty million of them. The cave was discovered only because the bats swarming out evening after evening looked from the distance like an enormous column of smoke, and would darken the sky for miles.

If the bats in their cave are roused during the day, the millions fly up in a dense confused mass. By means of infrared light this spectacle in the darkness has been filmed. The developed film was an immensely exciting sight: sometimes the picture was altogether black, so dense was the crowd of bats. Yet not a single collision took place.

In the millionfold rattle of this infernal din, how does every one of the creatures know which sound is the echo of its own call? How does its own voice stand out – the "cocktail-party effect"? We do not know. It is even more astonishing, considering that the bats become completely irritated by the comparatively simple jamming devices of the night moths. I shall have more to say about this presently.

By this phenomenal sense the bats change the darkness into daylight for themselves. Thus they dominate, almost without competition, a rich living space: the night with its abundance of buzzing and humming insects. One would think they must be really in paradise. But many insects have developed a number of tricks and counter-weapons which make things more difficult for the bats.

An insect's first counter-measure, as it were, is its own body; for while the bats are eating that, it effectively gags them and thus prevents their getting their bearings.

But the little brown bat has an answer to that. While chewing and swallowing it keeps open a sort of gap between its teeth, if at all possible, through which it can "whistle" almost as strongly as ever. But if the prey is too fat, then – according to Professor Anton Kolb[5] – a substitute transmitter goes into action: the creature emits sounding noises through the nose. Although these are shorter, of a deeper pitch, and only just half as loud, they at least save the bat from colliding blindly with trees or other bats. This works even better with big-eared and barbastel bats. When they want, they can produce as loud and piercing cries through their noses as through their mouths.

It has been left to another family, the horseshoe-bat, to make a virtue of necessity. These bats in all instances emit their probing sounds only through the nose. For this purpose they have transformed the outside of their olfactory organ into a trumpet, a megaphone. Although not perfectly round, but the shape of a horsehoe, it is outstandingly effective. With this parabolic mirror surrounding their nostrils like a moon crater, they focus the sound, as a radar device does with electro-magnetic waves, and beam it straight at a goal.

Professor F. P. Möhres[6] of Tübingen writes: "Just as to us humans a landscape becomes visible at night in the reflection of a searchlight, so to a bat its environment hidden in darkness becomes recognizable in the echo of its sound projector." Instead of light waves it is sound waves which convey to horseshoe-bats an absolutely graphic impression of their environment. Such are the facts, however much they sound like science fiction.

Furthermore, the horseshoe-bats apparently use a totally different principle of location from that of the bats mentioned earlier on. They do not send out volleys of ultrasonic bursts, but sequences of pure ultrasounds lasting much longer. They are probably the purest, least distorted, and most regular sounds produced in the whole animal world. Thus the large horseshoe-bat send technically flawless sonic waves on a frequency of 85 kilocycles and the lesser horseshoe-bat, which is common in Germany, on 110 kc. According to Professor Möhres, these fly-by-nights even recognize each other individually by slight differences in their wavelengths.

If the ultrasonic bursts of the bats with simple location mechanisms lasts only the thousandth part of a second, a tone transmitted by the horseshoe-bat goes on for thirty to fifty times as long. The horseshoe-bat does not transmit faster, the more it is interested in an object it has discovered, but it

Figure 60. The orientation cells of bats are very different. The oscillograph above shows the clicks of a bat with a simple location mechanism, which last only from one to two thousandths of a second; the one below shows the long-drawn-out, pure sounds of a horseshoe-bat, which last for a tenth of a second.

does keep the intervals between the long-drawn-out tones as short as it can. This means that it makes the tone sequence of its sounding noises as near as it can to a permanent tone. Naturally it cannot, however, completely suppress the need for breathing.

Here, then, the returning echo is nearly always drowned by the signal still being sounded. So the horseshoe-bats cannot possibly carry out a time-difference location like their clicking relations – and like all radar and sonar devices used by man.

From this comparison we see how tremendously fruitful it could be for human technology to explore the horseshoe-bat's sense of acoustic images. So far, however, we know practically nothing about it and can only venture a few guesses. The greater horseshoe-bat moves its external ears in time with the ultrasounds. With every call one ear moves forward and the other back, and this up to sixty times a second. Professor Möhres writes:

> The findings indicate that here it is not time-differences which are used, but differences of intensity in the echo rebounding from the environment. In this the movable ears work like direction finders. With the spreading of the ultrasonic waves in a straight line, the external ear can only catch the echo in full from a spot in its environment if its opening is turned precisely towards that spot.

The movements of the ears will then serve the purpose of scanning the environment point by point.

It is hard to imagine a photographic image being achieved in this manner. While the environment is systematically scanned – perhaps, as in television, line by line – the bat's brain may make every echo part of its image perception. That would be possible only if the auditory nerves worked with the nerves indicating the present position of the ears. The nerves of the sensory organs are so different from the position-indicators of parts of the body that it sounds improbable they should harmonize so exactly with each other. Yet such combinations are surprisingly widespread in nature, as anybody can discover by a simple experiment on himself.

Look at any fixed object. Now roll your eyeballs or shake your head several times. You never lose the correct impression that the object stays firmly in its place, although it is all the time being projected on different cells of the retina, so that you might expect it to look to you like a moving object. Now put a hand over one of your eyes, and with a finger of the other hand press lightly and rhythmically from the side against the eyeball of the open eye, moving it to and fro a little. At once the watched object seems to jump about like mad.

It would always do this at any movement of head or eyeball, if our brain did not provide a sort of clearing house taking into account such movements when receiving the messages of our visual nerves. By pressing our finger against the eyeball we have avoided the clearing house. As so often, the things we take completely for granted have become natural only because of an extraordinarily complex apparatus in the brain.

In a similar way the auditory nerves and ear-position indicators in a bat's brain might perhaps by direct cooperation cause a regular image to form. Morevoer, it may well be that in the nearer ranges of hearing a spatial impression forms, just as man can see things three-dimensionally up to three hundred yards away.

So the weapon of gagging has been ingeniously parried by the bats. Consequently, various insects defend themselves in another way: they make themselves inaudible. Besides not making any noise themselves, they are wrapped in "acoustic camouflage"; that is, in such a soft and sound-absorbing integument that it hardly reflects even the ultrasound of their enemies.

To catch buzzing, humming, whirring flies and mosquitoes, bats do not need their sound transmitters at all. They hear the flying sounds of their victims a long way off, locate them, and, hey presto! – the insect that gave itself away has had it.

In the same way bats locate the running noises of rustling beetles, spiders, and other running or crawling organisms. As Professor Kolb[7] has found out, half of the bats' food consists of these, while in spring nights when flying insects are rare, the bats feed exclusively on non-flying ones. Sure of their aim, the bats alight in the grass, moss or leaves, about three-eighths of an inch from a rustling insect and devour it on the spot. They even hover near a drooping leaf to collect insects which are crawling about on it.

Yet flies and mosquitoes multiply so boundlessly, that losses through bats threaten the survival of the species as little as do the windscreens of cars. But bats might be a serious threat to the survival of night moths as a species, had these not escaped the danger of complete extinction by developing absolutely noiseless flight – even in the realm of ultrasound.

According to investigations by the Berlin biophysicist Professor Heinrich Hertel,[8] this is what noiseless flight means in aeronautical terms. The moth has to avoid the forming of turbulence at the edges of its wings. This is a burning aerodynamic problem. Night moths solve it by fine fringes along the turbulent flow area. The little hairs are about two millimetres long and have a diameter of seven thousandths of a millimetre. This stops the moth from giving itself away.

But for a great many kinds of night moth this way of flying is still not silent enough. Thus the noctuidae (25,000 species

Figure 61. Fine fringes along the turbulent flow area of the moth's wing prevent the formation of air eddies which produce noise.

in the world), the geometridae (15,000 species), and articiidae or tiger-moths (6,000 species) have special ears for listening in to the enemy's transmitter. They turn the tables upon the bats, having arranged things in such a way that the bats give themselves away by their ultrasonic cries

The story of this magnificent discovery began on a warm summer's evening in 1956 when the American zoologist Kenneth D. Roeder[9, 10] gave a party for his friends on the verandah of his house. Chinese lanterns were lit, and soon there was a swarm of night moths whirring around them. During the evening one of the guests did a trick quite popular with wine-drinkers: he rubbed a damp cork round the rim of his glass, causing the familiar shrill ringing tone. The moths, which had been cheerfully fluttering round the lanterns, immediately dropped to the ground as if struck by apoplexy. Professor Roeder at first thought they had been paralysed or even killed by the shrill and nerve-racking sound. There are many examples of convulsions among animals caused by loud noises.

But instead, to the astonishment of all present, the moths were very much alive. They crawled along the ground for a bit, then started flying again. This strange behaviour has long been turned to account by butterfly collectors, to catch the nocturnal cousins of daylight butterflies with cork, wine-glass, and net or blanket; but as a result of the above incident Professor Roeder and his colleague Dr. Asher E. Treat decided to make a thorough scientific investigation of the phenomenon. After years of work at Tufts University in Medford, Massachusetts, their eventual results surpassed their wildest expectations.

The night moth's ultrasonic ears, they found, lie on both sides of the thorax near the waist and are an excellent model for achieving great effects by simple constructions. At first sight, indeed, it looks like the most primitive ear in the world. As Figure 62 shows, it consists of a tympanic membrane with an air-sac behind it, and a fine strand of tissue containing two nerve cells. These two auditory nerve cells send a fibre each to the moth's brain and a further receptor fibre each to the tympanic membrane – rather reminiscent of a record-player's needle. That is the whole thing, except for a third nerve, not reacting to sound, the significance of which we do not know. So here is an ear, for all practical purposes, with only two nerves.

In consequence, immediately a bat comes within a hundred feet of a flying night moth, the first highly sensitive nerve of the moth's ear for listening to the enemy registers the bat's

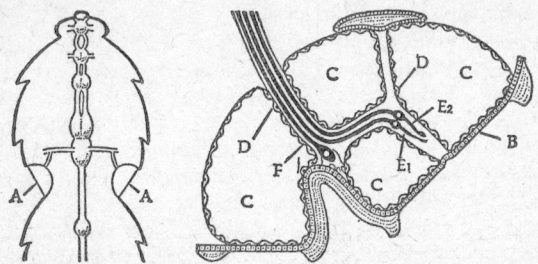

Figure 62. The moth's ultrasonic ears (A) are near the waist. Right: enlargement of an ear: the tympanic membrane (B), the air-sac (C), the supporting tissue (D) with the two auditory nerves (E1 and E2) and the nonacoustic nerves (F).

cries, and by a series of signals sends a preliminary alert to the brain. The moth at once swerves onto an avoidance course, in the opposite direction to the bat. If the bat is exactly below, the moth will even rise quite a way straight upwards.

Although it flies much more slowly than the bat, this alone often proves a life-saving manoeuvre, for by that time the bat's sonar has not yet located the moth. Because of the sound-absorbing padding of the moth's acoustic camouflage, the bat can only "pick up" its target within a range of twenty feet.

Wherever the bat flies, the moths a long way from its echo-location will scatter. If the bat flew as straight a course as the swallow, it would hardly ever catch a moth. So it has developed a counter-strategy, its reeling flight. This looks tremendously clumsy. The sight of a bat with its two umbrella wings seems a bit like Otto Lilienthal on his first attempts at flying. But that is an illusion. All the curious deviations are in fact different geometrical curves broken off abruptly by the bat to deceive its victims about the direction of its flight.

By this trick the bat often manages to approach within the sonar range of twenty feet. Then it receives the first echoes from the moth, at a sound intensity of thirty decibels, about equivalent to the noise of a fairly quiet car engine. Two things happen then: the bat starts heading straight for its victim, and in the moths' receiver ear the second auditory nerve cell begins to transmit: action stations!

Many night moths abruptly lay back their wings, dropping on the ground like bombs, as at Professor Roeder's party on the verandah. But since they thereby describe an exact ballistic curve, the bats after some practice learn to follow that curve, and their success in catching the dropping insects is about 50 per cent. So half the moths survive, which is very important for the survival of the species.

But there are night moths capable of even more sophisticated evasive action, and they are in a much better position. With the help of a searchlight Professor Roeder has filmed the night's operations hundreds of times. At the noise of an artificially produced ultrasound some moths doubled their tracks like hares to the left or right, others looped the loop, spun to the ground in close spirals, whizzed through the bat's "wake", or combined all these acrobatics to make the bats miss them. To contend with such air aces the bats have only one trick left: as soon as they get at least into wing's reach of a moth, they try to net it in the umbrella of a wing.

This is an amazing phenomenon for the neurophysiologist. Two primitive ears, with only four auditory nerves altogether, send messages to the tiny brain, where they are transformed into a different code and passed on to the wing muscles, causing the muscles at once to take the right action in any particular situation. It is one of the simplest cases of a connection between the construction of an animal's nervous system and the instinctive behaviour which that conditions.

All that concerns us, however, in the framework of this book, is how the night moths with such a simple hearing device can recognize whether the bats are approaching from left, right, front or rear, above or below.

With man too, direction hearing is still a great problem. But any radio amateur can shed some light on it by repeating an experiment of the physicist Dr. Hans Kietz[11] of Bremen.

The electric vibrations of a humming tone are conducted over two wires to a pair of earphones. Into one of the two conductors we connect a phase displacer – a sort of waiting room which delays the sounds on their way to the earphone by some hundred thousandths of a second. It must be possible to change the retarding time by turning a knob.

If the vibrations arrive simultaneously in both earphones, the experimenter is under the impression the sound is coming from straight ahead. The more the right side is now retarded, the farther left he imagines the sound is coming from. This is most extraordinary, for it proves that our brain can exactly register time differences of hundred thousandths of a second in the auditory sensations of the right and the left ear, can compare one with the other, and develop from that an idea of the direction.

It tells us nothing, of course, about how our perception is formed of directions in front and behind or above and below. We still do not know how the process works in man.

But to return to the moth. The insect first of all simplifies the problem a good deal by being tone-deaf. That means it cannot distinguish between tones of different pitch. So anything that produces a shrill noise in the night signifies "bat" to the moth, whether in fact a bat is crying with 110,000 vibrations a second or a man is producing the sound with only

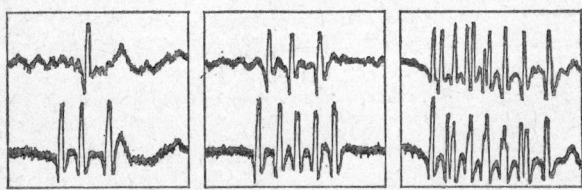

Figure 63. How the ears of a moth detect the approach of a bat. The bat is a good distance from the moth on the right. Every time the bat calls, the first auditory nerve in the right ear reacts with three impulses and the first auditory nerve in the left ear with only one discharge (left). The enemy comes nearer. This changes the firing ratio to five to three (centre). The bat darts on directly above the moth. So the auditory nerves on both sides fire with equal vigour (right).

The head of a common fly in close-up looks like a monster bristling with sensory organs: above, a compound eye, left, the antennae, below, the proboscis with the lobe at the end, the 'lip,' on which there are many sensillae of touch, taste and smell. There are many other sensory hairs to be seen on the rest of the head too.

The same creature seen twice in different ways. As seen by the
human eye (top) this American chameleon (*Anolis
carolinensis*) could easily be missed because of its good
camouflage. But as seen by the infra-red 'eye' below the
reptile shows up clearly, because the chlorophyll in the plant
reflects infra-red strongly, while the lizard absorbs this
radiation.

Alarm in the ant colony. A guard has noticed a robber beetle.
From its hind quarters it squirts the intruder with a scent
which then spreads out all around the beetle and draws a
concentration of aggressive ants on to it.

Top: The wasp, a brilliant piece of precision apparatus for flight and navigation.
Bottom: Limitation of army strength in the termite colony of *Trinervitermes havilandi*. With this species the soldiers have a 'glue-squirt' for a head. They stick enemies to the spot where they meet them. When the odour of soldiers in the nest exceeds a critical concentration, no more soldiers grow up.

The armoured face of a 'knight of the meadow'.

With the funnels of its two huge ears the big-eared bat can
'hear images' in a pitch-black night.

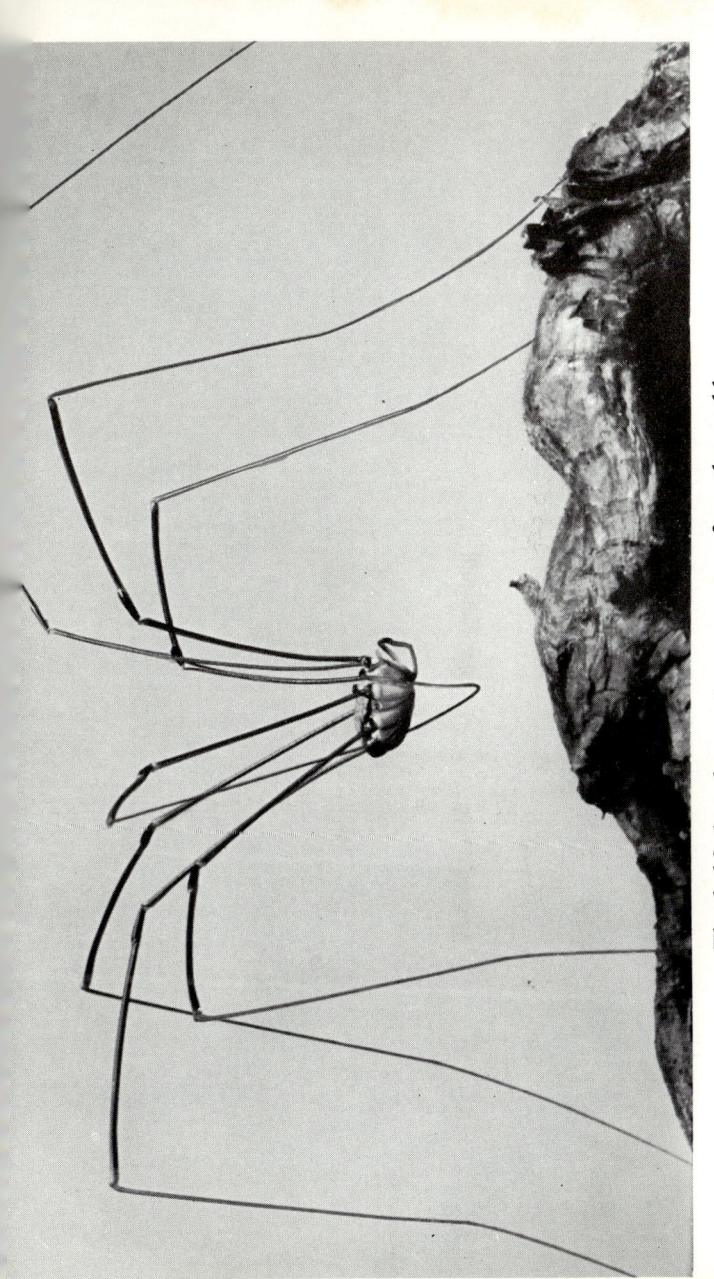

The daddy-long-legs, a strange creature from the world of tactile senses, an 'airship with legs'. It generally hunts at night and feeds on insects and spiders.

If a fly had human eyes it would see a garden spider like this — before being eaten. The spider, for all its eight eyes, is nearly blind, but understands perfectly the 'telegraphese' of the vibrations of its web.

10,000 vibrations a second. Both noises sound exactly the same to the moth.

Two of Professor Roeder's colleagues, Dr. Roger Payne and Joshua Wallman, found this out by recording from the moth's auditory nerves and listening in to the transmissions. They discovered that night moths, on the other hand, are very sensitive to differences in sound intensity. For instance, if a bat cries about twenty yards away on the right, auditory nerve No. 1 of the moth's right ear sends three impulses to the brain, but nerve No. 1 of the left ear sends only one.

This code alone, however, could lead to fatal errors. For the short, loud, and therefore near cry of a clicking bat produces the same series of impulses in the moth's nerve as the long-drawn-out and therefore distant cry of a sound-producing bat. The moth's nervous system takes into account the various methods of location of different bat species, in that loud noises not only trigger off more impulses, but also cause prompter reactions by the nerve cells, as can be seen in Figure 63. With these two forms of information combined, there can be no misunderstandings.

So the problem of right or left is solved. But how is the moth to tell above from below, and front from rear? Dr. Payne found the answer to the riddle: the wings are an indispensable hearing aid.

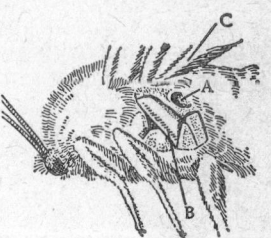

Figure 64. By the beat of its wings (C) the tiger-moth can cover or expose the ultrasonic ear (A) and the ultrasonic jamming device (B).

As Figure 64 shows, these very sound-absorbent membranes are placed a little in front of and a little above the ear. So the intensity of the sound received is also changing all the time with the beating position of the wings, which whirr up and down from thirty to forty times a second.

Roughly speaking, this is how it works: louder reception when wings are raised means "enemy approaching from in front and below"; when wings are lower, "danger threatens from in front and above."' No variation in sound intensity with the raising and lowering of the wings as they beat announces trouble from behind. When this scheme is combined with the differences of right and left in both ears, the moth obtains a characteristic sound image for every direction within its radius.

Everything seemed so simple at first: only four auditory nerves conduct all acoustic sensations to the brain. Now we add the indications by wing positions and are at once faced with a scheme of nerve signals of almost inconceivable complexity. Yet the insect's brain "understands" it faultlessly, and within a hundredth of a second transforms the signals into meaningful nerve commands for the control of the flying muscles. How complicated, then, the sense impressions must be in the brains of other creatures that set in motion thousands or even millions of nerves!

An electrical engineer might say: if such an excellent signalling system exists in the moth, it should be comparatively easy to connect further defence weapons to it. Many night moths of the tiger-moth family have in fact done this. At "action stations" they set an ultrasonic jamming device going, just like a bomber aircraft in the last war when it had been spotted by the enemy's radar apparatus.

Dr. Dorothy C. Dunning,[12] one of Professor Roeder's colleagues, has investigated the details. She constructed a feeder gun by means of which she hurled meal-worms high into the air. Tamed bats soon learned to catch these delicacies in their flight. But if a tape with an ultrasonic moth cry was played at the moment when the bat was about to snap at the worm, it would abruptly turn away and leave the worm untouched.

This is all the more surprising since bats do not let themselves be confused by any other ultrasounds, whether bat cries are imitated, permanent noise is arranged, or artificial jamming vibrations are invented. Yet the moth transmitter works on the simplest lines conceivable. As can be seen in Figure 64, at the juncture of the third pair of legs with the body there is on both sides a pliable grooved plate of chitin on top of a resonance box. When the moth alternately contracts and relaxes its leg muscles quickly, the plate starts swinging with ultrasonic vibrations, and within the reach of the bat's wavelength. Apparently the bat regards these sounds as a warning – but what is it being warned about?

Dorothy Dunning[13] discovered the following in 1968: North American moths of the genera *Arctiidae* and *Ctenuchidae* possess a very repulsive smell and taste. Bats spit them out immediately, if they swallow them by mistake. In order to save both the hunters and the victims from such unpleasantness, these moths have developed their own ultrasonic call as a distinct early warning signal, one might say, in order to advertise their lack of edibility.

A third moth genus, the *Pyrrharctica*, profits from this nightly propaganda battle in a clever way: it also emits ultrasonic sounds when approached by bats and is thus spared, although it does not have the slightest bad smell or taste and bats would undoubtedly welcome it as food.

Professor Hubert Frings,[14] however, who teaches at the University of Hawaii, plans to intervene in the acoustic war between bats and moths – for many night moths or their larvae are dreaded pests. Instead of contaminating large areas by poison, he hopes to protect regions threatened by pests with an ultrasonic curtain. Loudspeakers inaudible to man, "artificial bats", should make any moth crash and stop its ever rising again. Insecticides, as has been remarked, besides their dangerous side effects, have already become rather ineffective in their own function. It may well be possible that methods adapted by man to the insects' senses and behaviour patterns will prove more successful.

Underwater Sound Signals

In the early evening of a warm May day in 1942, the sirens howled in the Atlantic coastal stations of the United States Navy. As a protection against surprise attacks by German submarines, listening buoys had been positioned in Chesapeake Bay, the gateway to Washington and Baltimore. These buoys were to signal the propeller noises of invading enemy ships.

All hell seemed to have broken loose, as the buoys signalled underwater noises from all directions. Apparently a whole pack of U-boats was heading for the American ports. Destroyers and Coast Guard ships set out to plaster the approaches with depth charges. But no debris of any U-boats showed up, not even oil slicks. Instead the lacerated bodies of millions of fish were drifting on the surface next morning, their white bellies topmost.

About the same time something equally unexpected happened on America's Pacific coast. In front of its ports, to prevent a second Pearl Harbor, belts of acoustic mines had been positioned, designed to detonate at the noise of a propeller. All the mines exploded in a single night, yet not a trace of a Japanese ship was discovered. What had set them off? Not human sabotage, but the swim-bladder guitars of innumerable croaker fish. The fiasco occurred, in fact, as a consequence of the belief that fish were dumb and the underwater world a world of silence.

We have learned since that there are very few fish that do not make noises. In 1954 the author constructed for the Austrian underwater explorer and pioneer of amateur scuba-diving, Dr. Hans Hass,[15] an underwater microphone – that is, a hydrophone. When he let this down for the first time right amidst the teeming fish of a coral reef in the Caribbean, the loudspeaker on deck of his sailing ship *Xarifa* produced a strange medley of all manner of noises.

There was buzzing and whistling, knocking in a great variety of rhythms, rattling as if of heavy chains, booming, whipping, and sizzling like hot fat in a frying-pan. These were the fishes' cries. Perhaps they were cries of hunger, alarm signals, love songs, courting calls, or war-cries. We do not as yet know their purpose

A man abovewater, and in most cases even a diver underwater, hears nothing of all this, and without further knowledge would naturally assume that fish are dumb. Why do we not hear fish?

It is not as if, like bats or night moths, they made ultrasonic noises. On the contrary, fish noises usually have a very deep pitch, lying well within the range of human hearing. Our ears, however, are adapted only to hearing in the air. In water, different acoustic conditions prevail, for which we just haven't a suitable natural receiver. That can be supplied only by technology in the shape of the hydrophone.

How is this to be explained? The human ear has developed from the fish ear in the millions of centuries of the history of our earth. I shall presently be dealing in detail with the sensory organ proper, the inner ear: this is the cochlea (from the Latin for a snail-shell), which is filled with special fluids and has no air in it at all. In view of its origins it is not surprising that the cochlea is basically not a hearing apparatus for air sound but for liquid sound. Without the interaction of a transformer we might well, when diving, be able to hear the fish, but on dry land we should not even hear the language of other people.

In the form in which sound, funnelled by the external ear into the external auditory canal, sets the eardrum in motion, it is unable to stimulate our auditory nerves. That is why we have in the middle ear, in the so-called tympanic cavity, the indispensable air-liquid sound transducer; the hammer,

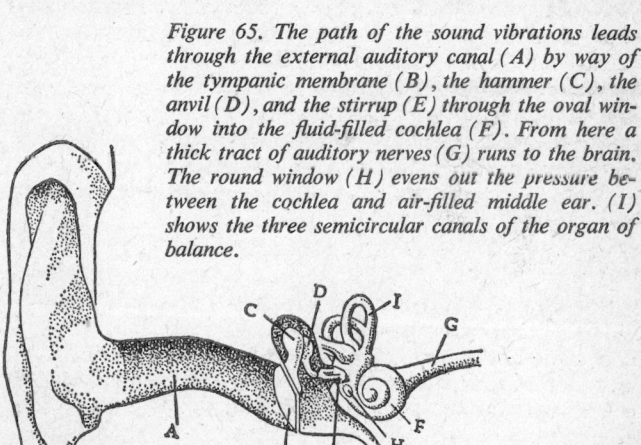

Figure 65. The path of the sound vibrations leads through the external auditory canal (A) by way of the tympanic membrane (B), the hammer (C), the anvil (D), and the stirrup (E) through the oval window into the fluid-filled cochlea (F). From here a thick tract of auditory nerves (G) runs to the brain. The round window (H) evens out the pressure between the cochlea and air-filled middle ear. (I) shows the three semicircular canals of the organ of balance.

attached by its handle to the eardrum; the anvil; and the stirrup.

These three small bones form a chain of levers passing on the sound vibrations. By this lever system the bones transform the vibration amplitudes of the eardrum and of the air sound, comparatively wide but weak, into those of the liquid sound, comparatively small but vigorous. The sound entrance, the eardrum, has an effective area of 85 square millimetres, whereas the sound exit of the middle ear transformer – that is, the waterproof oval window in the liquid-filled cochlea – takes up only 3·5 square millimetres. With a kind of stamping movement the stirrup concentrates its entire force on this, so that, amplified twenty times, it makes the liquid in the cochlea vibrate. The eardrum would be much too weak to do that.

With divers, things are just the other way round. The fish fill the water with their noise. But the sound vibrations of the water, intense though they are, have too little width of vibration and so can scarcely convey the same rhythm to the eardrum, which is not acoustically adapted to these vibrations. That is the only reason why fish seem silent to us. Sound in the water has to exceed a certain loudness[16] before we can hear it very faintly, even from the noisiest of underwater creatures.

One of these is the gurnard, which always reacts so grumpily when molested. It also seems to be an excellent weather prophet – its swim-bladder may serve as a barometer. At any rate, when fishermen in the Mediterranean hear it grunt, they make for their ports at once. The gurnard, as they are well aware, knows better than any human meteorologist when there is a thunderstorm brewing.

Hans Hass once got a gurnard to grunt into his hydrophone. It first investigated the apparatus with curiosity, then made a noise which in the loudspeaker sounded like a lion's roar. It may sound just as powerful to other fishes.

A terrific racket is produced by the toad-fish, ten inches long, living on the Atlantic coast of North and Central America. Dr. Hans Schneider[17] of Tübingen University says that during the mating season the male roars every thirty seconds like a foghorn. These sounds of spring are audible to the human ear, and Indian natives used to take them for spirit voices: they declared it was the sound of battle from the past winter, which had been frozen in the air and had now thawed out again. Some closely related species of fish give off shrill whistles, and are therefore called boatswain fish.

The pistol shrimp, two inches long, makes an even louder noise. Dr. Irenäus Eibl-Eibesfeldt[18] writes:

> With their claw, which is almost as long as their body, they stun the fish which is to provide their food. They point their claw at the fish like a pistol. When the shrimp has come near enough to its victim, the upturned claw finger snaps quickly shut; while a continuation of it presses water out of the counter joint, which squirts forward through a groove. The vibration is sometimes so strong as to splinter thick glass containers (as used for storage batteries) in which the shrimps were to be kept. Their activities filled the coral reef with rustling and crackling.

The din made by the croakers, the culprits which in 1942 fooled the American coastal defence in Chesapeake Bay, is a result of their enormous numbers. It is estimated that in this bay alone 300 million of these fish spawn and sound their rhythmical chug-chug-chug: always ten beats in a second and a half, then a pause of three to seven seconds, and so on. Professor Wynne-Edwards[19] believes this is a sort of census taking to limit propagation – like the population control by odours (see p. 156). For the more loudly a croaker fish hears the chug-chug-chug of other members of its species, the more it will restrict its production of offspring.

As has been said, there are hardly any fish that are completely silent; but their volubility differs a lot from species to species. Ocean fish make more noise than freshwater fish. Those in the tropics have a richer vocabulary than those swimming in northern waters. As with men, the fishes' voice gets deeper with increasing age – except for trout, which remain sopranos all their lives.

Even herring have a language, as was proved by scientists in the Rostock Institute for High Sea Fishing and Fish Processing. Schools of herrings signal in a sort of code by a soft chirping. Different signals have already become known which vary from each other by a chirping time of between 0·05 and 0·4 seconds. They mean: gather to form a school; establish vocal contact; beware predators; look out, change course.

Courting sea-horses click little serenades to their females. If the female is willing, she will softly click back. The boatswain fish fairly pipes up his sweetheart; the couples court each other like birds. But if a rival approaches, the whistling immediately changes into a deep grunt. Both adversaries spread their gill-covers wide, open their huge mouths and swim towards each other. In most cases one of them is intimidated by the mere grunt of the other and leaves the female to him without any struggle: it is a battle of noise with no bloodshed.

Anemone fish[20] (*Amphiprion bicinctus*) too try to scare away rivals of the same species by threatening noises if these lay claim to their anemones. If the threats fail to work, a fight ensues with ramming thrusts, fin blows, and tugging by the mouth. The interesting thing is that *if* the duel takes place in an aquarium and the loser cannot escape, he still has a means of protecting himself from further assaults by the stronger one: he utters humble, appeasing sounds, a protracted quacking noise. Then the victor leaves him in peace.

Very few fish use their mouths for making noises. Triggerfish, moon-fish, and soldier-fish produce creaking, squealing, and screeching noises by rubbing together their projecting incisor teeth. These are grooved like a gramophone record and produce very special noises. But most fish are percussionists. They drum on their swim-bladders with their muscles and ligaments, or they use ligaments extended over the sounding box of the swim-bladder as a string or plucking instrument.

Their process of hearing seems to us equally strange. We shall look in vain for ears on the surface of their bodies. They are deeply embedded in the fish very close to the brain. They have developed from part of a sense organ that has nothing at all to do with the normal perception of sound, and registers only slow pressure fluctuations of the water, infrasound, as it were, too deep for us to hear. This is the lateral-line organ.

How Acoustic Sense Organs Work

Hans Hass once made a dangerous experiment in the Caribbean. He had been wondering why, a few seconds after he had harpooned a fish, a shark previously invisible would nearly always come darting along to contest his catch. How did the shark know there was something to be had there? It could not see, for it was far outside sighting distance. Nor could it have smelt blood, because odours do not spread with lightning speed. Could it have heard something?

Dr. Hass caught a large bass, but this time with a sling. The fish at once started struggling wildly, and as he had hoped, a shark turned up in a matter of seconds. The same moment he loosened the sling, so that the bass calmed down and swam off as if nothing had happened. Thereupon the shark merely looked a bit puzzled, and without molesting the bass swam around a few times and disappeared.

Obviously the struggling bass with its wildly flapping fins had created an underwater rumpus, inaudible to the unaided human ear, which had brought the shark to the spot rather like a policeman summoned to a disturbance of the peace. When the shark arrived, the noise that had triggered off its urge for food ceased, and it no longer had any incentive for action.

Incidentally, a hungry shark hunts in a different manner. From great depths it will suddenly dart upwards, only just above the steep precipice of a coral reef, taking the fish by surprise and cutting them off from the shelter of the reef. Then it chases them into the open sea and catches them there. In regions where fish are plentiful, sharks never go hungry. But besides that, the noise of a struggle and the scent of blood arouse in them an irresistible urge to destroy the injured and diseased. That is why they also attack bathers and survivors of a shipwreck. But to a skin-diver who keeps still, their reaction is generally the same as to the bass freed from its sling.

This explains how acoustic stimuli release the urge, but not how the shark can so quickly find the invisible source of the sound. For the ear embedded right in the body is most unsuitable for telling directions. Consequently, fish use a different sense for locating centres of commotion in the water: their lateral-line organs.

On many fish one can see a line on each side of the head, leading right to the tail; it looks as if their scales were joined together along this line. Some people fond of eating fish even believe that the lines are there to show where best to cut the fish up! But in fact this outward mark conceals a strange construction of nature's which Professor Sven Dijkgraaf[21] of Utrecht University has investigated in the shark (things are a little different with other fish).

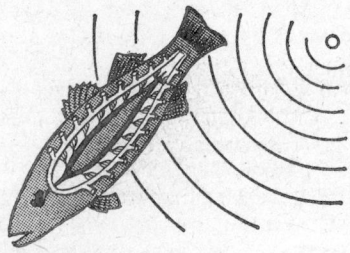

Figure 66. The lateral-line organ of fish (organ enlarged in drawing out of relative scale) registers by countless "portholes" (here shown much simplified) a wave of water pressure running along the body. Nerve endings (black) in the system of tubes signal the pressure waves to the brain, which can tell from the time differences where the waves are coming from.

On each side there is a thin tube embedded in the skin, running beneath the lateral line, which is filled with a special fluid; the two tubes join up in the head. There are many short blind-alley side-channels leading out of these veinlike tubes, sealed off against the surrounding water at their far end only by a very thin membrane: countless eardrums arranged as a row of microscopically small portholes.

So the vibration waves of the water spread in part to the fluid in the tubing system. But here there are fine sensory cilia all along the inside of the tubes, which are agitated by the wave movement like cornstalks in a gust of wind, and which report this stimulus to the fish's brain. They are virtually tactile hairs which through the hydraulic apparatus carry out a completely new task. By an attachment, then, the close or contact tactile sense is turned into a distant tactile or vibration sense, which in turn represents the preliminary stage of hearing.

By the time difference of a pressure wave arriving at the front or rear, left or right, the fish can tell very exactly which direction the wave is coming from. With freshwater fish this is first applied in recognizing the direction of the current in river or stream; with coastal fish in recognizing the direction of the tidal current (see p. 79). The next step is to register swimming fish, which, like our ships, create pressure waves noticeable over a wide distance. These fish may be members of the school in the immediate neighbourhood, but they may also be prey or enemies a little farther away. Finally, the fish can receive echoes of their own pressure waves caused by meeting obstacles while swimming.

The sense of vibration is so finely developed that it can even replace the eyes. This is proved by the blind fish living in the eternal darkness of lightless caves and subterranean watercourses. Lateral-line organs and noses are perfectly sufficient for them, although these two senses are no more efficient in blind fish than in those that can see.

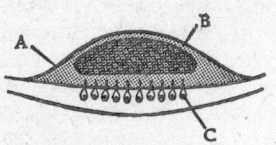

Figure 67. A "primitive structure" of the ear: the otolith of the ray. An otolith (B) is embedded in a gelatinous mass (A). When set in motion by sound waves, it stimulates rhythmically the sensory hair-cells beneath it (C).

The method by which nature has developed ears from the lateral-line organ is very striking.[22] At first otoliths appeared in the place where the main tube of the lateral-line organ leads over the head. These are made to vibrate by sound waves and thereby exert a rhythmic stimulus on a pad of sensory cilia lying beneath them. With every vibratory motion every sensory cell fires an impulse to the brain. An early model of this type of ear, as we find it in the ray (one of the oldest fishes in evolutionary history), because of its intertia reacts only to a very deep bass. This structure cannot follow sounds above 120 vibrations a second; to the ray, human language is inaudible, ultrasonic.

Members of the carp family, salmonids, groundling, knife-fish and bull-heads, are much better off. With them the swim-bladder is connected to the otolith by a chain of auditory ossicles (small bones). The top limit of hearing in the minnow is between 5,000 and 7,000 vibrations, and with the dwarf catfish it even reaches 13,000.

But the snag is that these fish cannot distinguish pitch really well except between 400 and 800 cycles a second, because their hearing organs consist only of two inarticulate pads of sensory cells, one for the deeper and one for the higher tones. So the fish hear only a uniform deep tone in the entire lower spectrum, a short "scale" above it, and in the higher spectrum again everything is a uniform high tone.

We can imagine this with some difficulty. For man's range of hearing does not, as is generally believed, extend from 16 to 20,000 cycles per second. Dr. Hans Kietz and Dr. Claus Timm (ear, nose and throat specialist at Lübeck)[23] have proved that we can still hear so-called ultrasound quite well provided the transmitter has direct contact with our skull bones without any air intervening. But then we register any number of vibrations from 20,000 to 176,000 – the upper limit Dr. Timm has checked – as of exactly the same pitch. Although this "ultra-range" comprises well over three octaves, to us all the 36 tones physically possible seem stereotyped and exactly like the highest tone we can hear in the normal way when it reaches us through the air.

Nature had therefore to make a particularly brilliant invention to endow birds, mammals, and men with the gift of hearing so many different tones. This invention is the basilar membrane of the inner ear.

What does the inner ear in fact do? Let us assume we hear the three octaves above middle C, and then the stirrup vibrates 2,048 times per second on the fluid column of the cochlea. But it sounds different if we hear the same tone from a violin of a flute. The reason is that with every instrument the fundamental tone of 2,048 cycles per second is overlaid by several harmonic vibrations (4,096, 8,192, and 16,384 cycles per second) of a characteristic volume, so that a pure tone changes into a colourful sound.

Let us draw up a fundamental vibration with its first and second harmonic overtones. As Figure 68 shows, if we then shift the phase relationship, we get a totally different course

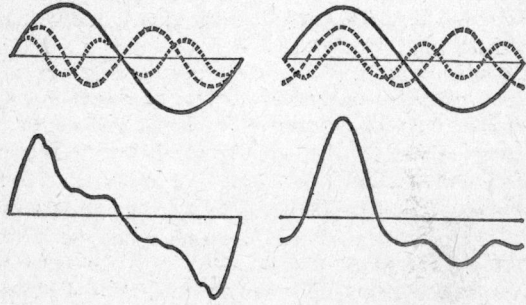

Figure 68. Two sound waves made up of the same fundamental and two harmonics, but in different phase relationship.

of the combined curve; that is, of the pressure course which is apparently effective in the final analysis. When looking at both curves, one might suppose that in the two cases we should hear two sounds that are just as different. But in fact our ear does not notice any difference.

What it amounts to is this. While our eardrum, the auditory ossicles and the fluid in the cochlea vibrate according to the course of the combined curve (as graphically represented by the grooves of a gramophone record), there must be a mechanism somewhere in the ear that splits up the resultant curve into its constituents; a series of absolutely pure tones (sinusoidal vibrations), a fundamental tone, and several harmonic tones. In other words, this mechanism for splitting up the sound must arrive at the same results in both of the cases in Figure 68, regardless of the phase relations. It must not, as our eye does, treat the two curves as two different things.

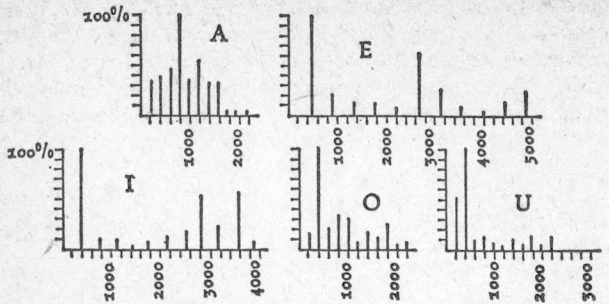

Figure 69. The vowels of human language are a mixture of several pure tones. Here we see their composition (horizontal scale, measured in vibrations per second), and loudness (vertical scale, in the percentage of the loudest part).

In theory we had best imagine it like this: somewhere in the ear there are "strings" as in a piano. When the vibration of Figure 68 hit these, then in every case the same three "keys are struck": the fundamental tone, the first harmonic and the second harmonic.

People with technological knowledge might say that, like a tongue frequency meter, the ear must be capable of the harmonic or Fourier analysis. In practice, the picture is of course much more complicated, since every sound contains a number of harmonic vibrations. Figure 69, for instance, shows which tones covibrate with the vowels of human language and at what volume. This becomes overwhelming when we listen to a symphony orchestra.

On the basis of theoretical reflections alone, the physicist Hermann von Helmholtz had claimed that we *must* have such a device for splitting up sounds in our ears. But for a long time there was no certainty of where this device was situated and how it worked. To find the device, we need to take a closer look at the cochlea.

At first sight this bone structure has indeed a striking resemblance to a snail-shell. In man it measures about 0·0305 cubic inches and has two and a half spirals of a total length of 38 millimetres. But the decisive difference from the snail-shell is that the spiral is subdivided into two parallel canals, the vestibular canal and the tympanic canal, which are separated by a wall of a very special kind. At the tip of the cochlea, as Figure 70 shows, the two canals join into one.

This is the process. Through the sealed oval window the vibrating stirrup sets only the fluid column of the vestibular

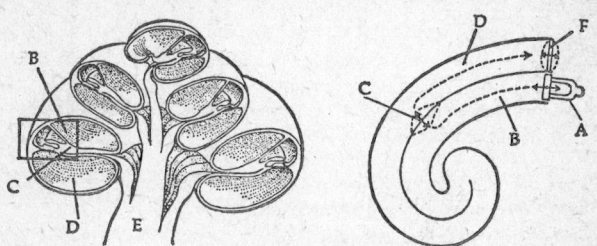

Figure 70. Here the sound is divided into its elements: cross section of the cochlea. The stirrup (A) transfers the sound vibrations to the fluid column of the vestibular canal (B). A part of the basilar membrane (C), corresponding to the number of vibrations, between the vestibular canal and the tympanic canal (D) starts vibrating. This vibration is picked up by nerve cells and signalled to the brain by way of the tract of auditory nerves (E). The adjustment of pressure of the vibrating fluid column to the air-filled middle ear is produced by the round window (F).

canal into motion. The impact winds its way up to the tip of the cochlea, where it is transferred to the tympanic canal, winds its way down again, and presses against the membrane of the round window, which produces an adjustment of pressure to the air-filled middle ear.

Professor George von Békésy, formerly at Harvard, now in Hawaii, was awarded the Nobel Prize for elucidating something that happens during the process. When, for instance, the ear receives a perfectly pure tone, the basilar membrane between the vestibular canal and the tympanic canal starts vibrating at a particular place in the spiritual tube. The higher the tone, the nearer the vibration spot is to entrance and exit; the deeper the tone, the farther the mechanical excitation shifts towards the tip of the cochlea.

In any case, we see here, in the vibrating spots of the basilar membrane, the device that analyses the sound. With a sound

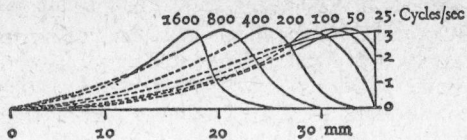

Figure 71. These graphs show which sections of the basilar membrane in the cochlea start vibrating when stimulated by pure tones of various pitches. The horizontal scale gives the distance in millimetres from the entry; that is, the oval window. The vertical scale indicates the strength of the vibration amplitude.

as in Figure 68 three different spots vibrate; with the vowel A, according to Figure 69, eleven different spots; and with a symphony orchestra the whole wall vibrates – more in one spot, less in another; and this changes with every fraction of a second. So in a most prosaic manner, the most beautiful harmony is physically analysed and codified into the electrical volleys of innumerable auditory nerves. Only in our brain is it reconstructed into a particular sense impression we call music; and the same with what we call language or noise.

The way the auditory nerves probe the vibrations is very

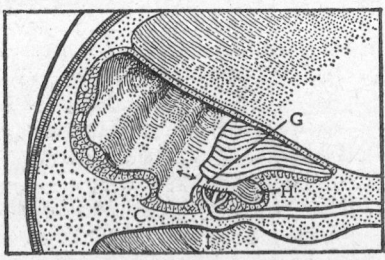

Figure 72. This enlargement of the section marked in Figure 70 shows the decisive details of how the auditory nerves are excited. The point where the strongest vibration is imparted to the basilar membrane (C) by the sound vibrations is above the vertical double arrow. Auditory fluid is thereby rhythmically pressed out through a narrow passage (G) from the bay-shaped space (H) and back. The strong current to and fro in the narrow channel passes over a pad of sensory hair cells, the organ of Corti, and stimulates them.

clear evidence that the internal ear derives from the lateral organ of fish, as is made plain in Figure 72. In a narrow passage, with a regular squeezing movement, the vibrations of the wall are at first transformed into vibrations of the fluid; and these in turn stimulate the groups of auditory hair cells. Without this wonderful sound-dissection device of the cochlea, birds, mammals, and men would not be able to experience their rich worlds of sound.

Moreover, the inner ear is like a fish's lateral line organ in another way too. The nerve cell reacts by a discharge to every up and down of the vibration. But the firing speed of a nerve

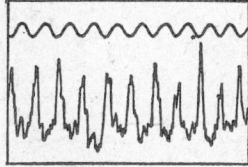

Figure 73. This excitation pattern (lower graph) is sent by auditory nerves of the budgerigar when they are stimulated by a pure sinusoid tone of 1,015 vibrations per second (upper graph). The nerve cell reacts with a discharge to every up and down of the sound wave.

cell cannot be indefinitely increased (with the four octaves above middle C it would have to send 4,096 impulses a second), so nature applies the principle of division of labour. Out of a group of closely adjoining nerve cells only one reacts at any given time, while the others are inert. At the next vibration it is the turn of another nerve, and so on.

This technique of communication is above all suitable for signalling pitch. Information about loudness comes off rather badly, although it is possible to a certain extent: the louder the sound, the more auditory nerves take part simultaneously in signalling one and the same vibration. But when we consider that man can distinguish over three hundred and fifty degrees of loudness from a soft whisper to the thunder of a jet plane, it is obvious that this signalling technique, with which the lower vertebrates and the birds probably have to make do, is inadequate for men and mammals. In fact, only about a third of the auditory nerves in our inner ear work in the way I have just described. The reaction of all the others is not synchronized with the stimulus. They are highly modulated gauges of loudness.

There is one thing in all this that seems strange. One might conclude from Figure 71 that the human ear can differentiate only very roughly between pitches. The same impression is formed when we investigate the sensitivity range of the auditory nerves. If we look at those at the point in the spiral passage which has its maximum vibrations at 1,100 per second, we find that they respond nearly equally well to vibration figures of 1,000 or 1,200.

But in fact our ear has a good deal better analytic powers. For we can distinguish even 100·0 from 100·1, or 3,000 from 3,005 vibrations a second, by the pitch. The explanation of this amazing fact is simple in principle, but immensely complicated in its details: the crudity of the acoustic device in the cochlea and the reaction range of the nerves must be compensated for by utter precision in the statistical processing of the signals in the nervous system.

"Hearing may be regarded as a sort of signal receiver with subsequent data processing," said Professor E. Zwicker[24] of Stuttgart. We might conceive of a circuit that takes note only of maximum excitation, suppressing anything that vibrates with it, but only in so far as it really does join in the vibration; for any excitation, however delicate, by another tone of the harmony must of course be allowed its full effect.

So here we have again come up against the frontiers of the unknown.

All this is the more magnificent because the nerve circuit enabling us to hear which direction a sound is coming from is also linked to it. Figure 74 shows a diagram of the nerve connection from the internal ear to the brain, as drawn up by Mark R. Rosenzweig,[25] Professor of Psychology at the University of California.

If the source of sound is exactly 90 degrees to one's right, the air vibration will reach the right ear one two-thousandth of a second earlier than the left one. If the sound comes from only 5 degrees to the right, that is, almost from straight ahead, the time difference amounts to only one twenty-five-thousandth of a second. But the nerve signal needs a hundredth of a second to run from the ear to the brain; that is, 20 times or 250 times as long. How should such sluggish nerves be able to register such minimal time differences? For decades it was thought impossible.

Yet they do. Professor Rosenzweig has traced the auditory nerves in anaesthetized cats electronically and found out that the time comparison concerning left and right does not take place in the brain, but much earlier in special nerve ganglions.

Figure 74 shows a number of such switchboards. According to Professor Rosenzweig's theory, in each of these a stimulus arriving earlier inhibits the passing on of another stimulus which arrives some ten-thousandths of a second later. It is a sort of relay race in which the runners – that is, the electric signals – are all equally fast, but have to cover distances of different length. They are sent along diversions or take short-cuts. They change over from the right nerve bundle to the left and sometimes back again. They divide and join up with the most varied partners.

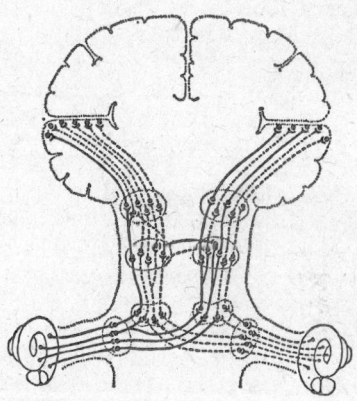

Figure 74. The nexus of auditory nerves on the way from the cochleae to the brain is reminiscent of the wiring of an electronic computer. Here, time differences of ten thousandths of seconds are converted into different levels of loudness, which in turn are used as a protractor for direction hearing. The much simplified diagram shows the course of some nerve tracts from the left ear (continuous lines), the right ear (dotted lines), and of combinations of left and right (dash and dot lines).

This produces the following result: if the source of sound is to the right, a considerably stronger electric stimulus is generated in the left hemisphere of the brain than in the right. So the minimal difference in time is converted into a difference in loudness that does not really exist. The more loudly a sound is registered by the left, and the more softly it is registered simultaneously by the right region of the brain, the farther to the right it originates.

A sound coming from the right does, of course, also strike the right ear more forcibly than the left one, which lies in the acoustic shadow of the head. And since most of the auditory nerves of the right ear change over to the left hemisphere of the brain (see Figure 74), this is in any case stimulated more strongly. But the real difference in loudness is so negligible that in itself it does not suffice for location. So it gives some support, though only a little, to the ostensible difference in loudness into which the time difference has been converted. This is the secret of direction-hearing.

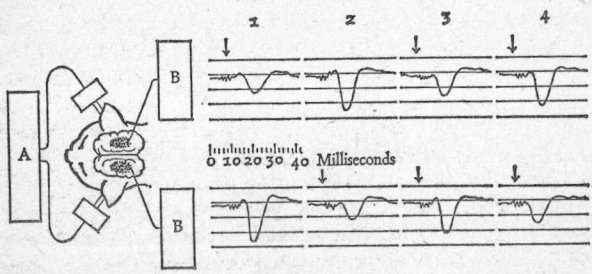

Figure 75. How the auditory region in the brain of an anaesthetized cat reacts to four different sound stimuli. While a tone generator (A) sends clicking noises into the ears, the brain's activity is monitored with electrodes and made visible with two recording devices. (B). Example 1: on a click only in the right ear the left hemisphere of the brain reacts more strongly than the right. Example 2: on a click only in the left ear it is the reverse. Example 3: if equally loud clicks are directed to both ears, but the one on the right ear two ten-thousandths of a second before the one on the left, the graph of the brain impulses resembles that in Example 1 (sound only in the right ear). Example 2: if the click hits the left ear two ten-thousandths of a second before the right, the opposite effect is produced.

Here again we marvel at the wonderful methods with which our nervous system works, at the simple principles by which it masters things long regarded as impossible, and at the incredibly complicated techniques by which it puts these principles into practice. When we consider the process of growth by which all these countless nerve connections are formed, compared to which the wiring of an electronic brain is child's play, we can scarcely avoid recognizing the existence of some higher force.

The Problem of Recognizing Language

The dream of bosses all over the world is a "robot secretary," which would cut out constant troubles with personnel. You would simply switch on a perfectly functioning machine to take dictation: as soon as you had spoken the text into a microphone, a fully automatic typewriter would hammer the letters faultlessly onto the writing paper.

There is a nervous mechanism in our ear and brain which can pick out individual sound signals from the human language and recognize their meaning. It might well be desirable to have an electronic data-processing plant that could do the same, but the difficulties of producing such a plan seem more or less insurmountable.

They have been outlined by Dr. Werner Endres[26] of the Research Laboratory of the Central Institute for Telecommunications in Darmstadt. The apparatus would have to be capable, first of splitting up the continuous flow of speech into separate words and writing these down in orthographically correct form; then of recognizing the syntax (from the inflection of voice among other things) and putting in the punctuation marks. It would, in fact, need to have complete mastery of the language to be processed; and whether man will ever succeed in producing such an apparatus remains highly doubtful.

Any disappointment at our capacity in this field should be balanced, however, by admiration for the "technical" efficiency of the hearing and speech regions of our brain, which nature has created. Now that we are trying to achieve the same results, we appreciate the full greatness and intricacy of the processes necessary for recognizing language.

In any case engineers in a number of institutes in America and Europe have not given up trying, though for the moment limiting their objectives. For instance, they ask the person dictating to the automatic typist to speak very distinctly, make long pauses between two words, indicate the punctuation marks, and be content with a phonetic reproduction. Under these conditions there can be no question of the robot's mastering the language.

A phonetic alphabet consists of about forty to fifty signs. The most important of these are reproduced in Figure 76. So for the sake of the inadequate machine we should have to learn a new script and completely change our spelling. It would be much simpler, and would greatly reduce the possibility of spelling mistakes but the break with tradition would strike the guardians of our cultural heritage as barbaric. Yet who knows what changes will be forced upon us in an age of progressive automation?

Even if the writing robot is simplified into an orthographic idiot, many other difficulties remain. It is by no means the case, unfortunately, that each letter of the phonetic alphabet has an absolutely characteristic, invariable image of vibrations (as one might be inclined to conclude from Figure 69), which would be ideal for automatic processing. "It has become evident, on the contrary," writes Dr. Endres, "that in natural language the individual sounds continually melt into each other, and that the vibration course of every single sound is influenced not only by the transitions from and to the immediately adjoining sounds, but also by sounds before or after these."

For spoken words are not, like written ones, composed from standardized individual parts. The sound pattern corresponding to a spoken word or phrase is rather like a letter written with a lot of different flourishes which depend on the letters before and after it. One of the things we do naturally when registering speech is to pick out which parts are important for understanding what is said, and which are mere flourishes. We cannot yet get a machine to make this distinction.

Voiced sounds	Diphthongs		Voiceless sounds	
	au	taut		
	ou	out		
	i	ice		
	Labial and bilabial sounds		*Fricatives and Sibilants*	
Vowels				
a hat	m	moon	f	fish
a hard	n	noon	s	sun
e bed	ng	song	sh	shot
e read	y	year	j	jump
i it	l	lot	h	hat
o hot	r	red		
o note	w	wood		
u rub				
u tune				
	Fricatives		*Plosives*	
	v	very	p	pot
	z	zoo	t	top
			c	cat
			q	queen
	Plosives			
	b	ball		
	d	dot		
	g	good		

Figure 76. Survey of most important sounds in English language.

So in constructing apparatus for recognizing speech, we must for the time being put up with further restrictions. At the moment several research workers are trying to seize on some rough characteristics in the sound image of speech, on the basis of which an electronic device could automatically distinguish at least a dozen words from each other. At the time this book is being written, therefore, the vocabulary of these machines is still absurdly small.

The most original of all speech recognition machines has

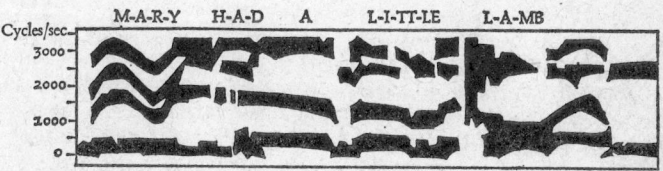

Figure 77. An example of a voice print. This shows the different kinds of vibration from 0 to 3,500 cycles per second, which make up the separate sounds in the sentence, "Mary had a little lamb."

been developed by W. C. Dersch[27] at IBM. This can recognize
and print sixteen words dictated into a microphone, no matter
whether the speaker is an adult (with bass voice) or a child,
a man, or a woman. The "words" are the numbers zero to
nine, and the instructions plus, minus, subtotal, total, false,
and off. But the machine still makes a mistake every seven-
teenth word. As it is no bigger than a shoebox, "shoebox" is
the name by which it has come to be known.

Similar machines have been built by Dr. H. G. Tillmann[28]
at Bonn University, and by H. Kusch.[29] There is also reported
to be a scheme for having astronauts control their steering
gear by spoken orders while they are roaming through space.

Future development will probably continue in two directions. One is a plan to use speech automata with small
vocabularies as remote controls for machines over calling
distance, and as acoustic feeding machines for electronic
computers. Secondly, scientists are trying further to improve
the systems of speech recognition and to increase the machines' vocabulary up to about one thousand words. For
simple office language this would be quite adequate.

A certain parallel can be seen between the development of
these speech recognition instruments and the evolution of
the hearing and speech region of our brain. There are surely
animals that are totally unable to grasp various sound sequences of the human language. Many song-birds obviously
have a power of comprehension that is more or less confined
to their own species' repertory of tunes. Mocking birds and
other songbirds, however, prove by their imitations that they
also pick up sound patterns foreign to their species. Parrots,
ravens, and mynah birds even pick up a vocabulary of human
words, without much understanding, while many mammals,
including chimpanzees, dogs, and elephants, learn to recognize enough human words to obey our orders.

When we consider that this power of comprehension is the
precondition of speech, an essential feature of our humanity,
we can understand what enormous demands an employer is
really making when he wants an electronic typist for his
office. Such a device would be not far removed from a human
robot. So perhaps it is a blessing, after all, that we cannot yet
make one.

Chapter Six

Senses of Touch and Vibration

The Vibration Sense

"TOWARDS ten o'clock in the morning screaming swarms of sea birds darkened the sky above the town of Concepción on the Pacific Coast of Chile. At 11.30 the dogs fled out of the houses. Ten minutes later an earthquake destroyed the town."

This report from the year 1835 hails from the British meteorologist and admiral, Robert Fitzroy. We have been told ever since after every earthquake that before it happened horses had been trembling, dogs barking in a peculiar way, chickens running about in a panic. It was the same story with the disasters around 1960 in Agadir, Skopje, Chile, and Alaska. In the villages on the slopes of Mount Etna the peasants keep cats because they believe these animals can anticipate volcanic eruptions: when all the cats leave the house at once, men will rush out after them.

Are all these old wives' tales, or is there a grain of truth in them? At the present time, when many geophysicists are trying to work out a foundation for forecasting earthquakes the question is of special interest. It is at least conceivable that animals possess a sense for the physical events that lead up to an earthquake.

So in 1960 for the first time a scientist followed up this strange phenomenon; it was the German zoologist Dr. Ernst

Kilian,[1] who teaches at the University of Valdivia in Chile. He hoped to discover whether animals can be used to forecast earthquakes and which portents they can sense.

Objective observations are very difficult because of the primitive and panic fear of earth tremors which is felt by man as well. After the main quake shook the town of Valdivia, Dr. Kilian could not even tell if he had been able to keep his balance or had been thrown to the ground several times. But in the course of the eleven big tremors, and over a hundred light ones that followed in the first three days after the quake of 1960, it was possible to do some good research.

The stud-horses in the University's stables started neighing and trembling five seconds before every tremor. A ring-necked pheasant announced every quiver by loud gaggling ten seconds before men could feel it. Very slight tremors, which did not worry Dr. Kilian at all, affected dogs so much that for several minutes their whining was heard all over the town. Other animals, however – sheep, chickens, and three captive pumas – showed themselves decidedly thick-skinned.

So it seems very possible that animals with an extremely sensitive vibration sense receive "warnings" through advance tremors almost imperceptible to man, which sometimes precede the earthquake proper by several days.

But this is far from explaining the whole secret; for the earth is shaken by 150,000 light tremors every year, but only about twenty are preliminary to an earthquake. Nor do animals show the slightest emotional reaction to light vibrations of the earth's crust, unless they have been disturbed by a strong tremor shortly before.

There are other things besides slight tremors that herald an earthquake, such as storms in the earth's magnetic field; also, the bursting of the crust from internal pressure is sometimes triggered by extremely high barometric air pressure. Both natural phenomena ought to be perceptible to animals with a magnetic sense and an internal barometer. But here again, in not more than one in a hundred thousand cases will these phenomena be followed by an earthquake.

Dr. Kilian believes it possible, however, that a certain vibration pattern of the advance tremors, or a particular coincidence of these and/or other factors still unknown, characterize the prelude to an earthquake; the corresponding combination of different sense impressions might then have the effect of alerting the fear centre in the animal's brain.

This has not yet been proved, and may seem very strange; but in view of all that has been discovered about animal senses in recent years, the "more things in heaven and earth" principle must apply. And certainly some animals can interpret shocks through their vibration sense in the most astonishing way.

For instance, the male Siamese fighting fish (*Betta splendens*) which looks after its brood, always has a lot of trouble with them. Two or three days after being hatched, they court danger by leaving the nest to have a look at the world of their South Asiatic inland lake. The first thing the males tries is to take the runaways back to the nesting place in his mouth. Because of the great number of offspring, this soon becomes more difficult than looking after a bag of fleas.

Consequently, the harassed father acts as follows (in the description of Dr. Wolfdietrich Kühme):[2]

> It floats at an incline with its mouth just below the water surface and makes its pectoral fins tremble violently. The young fish up to sixteen inches away take their bearings on the centre of the vibrations and arrive there in batches. The longer the male keeps on trembling (it can do so up to five minutes), the more young fish collect beside its head. With a few sideways motions of the head it sucks them in, repeats the procedure in various places and spits out all the young fish at the nest location. About five days after hatching collecting activities slacken very quickly. The young fish swim more than an inch below the surface and no longer register their father's quivering.

The length of time for which the vibrating goes on is apparently the sole criterion of whether the thing making waves is just an object that has been dropped in the water or father calling.

Finer differentiations are lacking also in mating behaviour. For (as described by Lorus and Margery Milne[3], many fish do not touch each other in mating. Whether the fish are two-inch stickleback or forty-inch salmon, the male only has to start quivering violently near the female to make it deposit its eggs.

But the same reaction can be caused by man if he sticks an oar into the water and makes it vibrate near a salmon female which is ready to deposit its eggs. Animals' "ideas" of their mate are very different from man's. With many fish these include among other things a quivering excitement, no matter where it comes from.

The vibration sense is much more differentiated with web-weaving spiders. One may say without exaggeration that the web serves also as a telegraph line by which various messages can be received. All web spiders live in a tactile world. Their visual sense is little developed and plays a very minor part in their lives. If an opaque mask is stuck over their eight eyes, they live, mate, feed, and flee just as well as sighted spiders. But how do they know whether it is prey that has fallen into their toils or an enemy, whether a suitor is approaching or a young spider trying to make off? These questions have been investigated by Dr. Erwin Tretzel[4] of the University of Erlangen.

The male of the garden spider "phones" his chosen bride. He attaches a thread to her web, which he plucks in a certain rhythm. Related species cling to such a thread in order to make it vibrate by jerky movements. By the way the female reacts and by what kind of vibrations she sends out in her turn, the male knows how great the danger is of being eaten by his bride, or whether he can risk mating.

Even more interesting is the communication between a mother and her young. With the funnel spider (*Coelotes terrestris*), which weaves a large, fine-meshed web spread out horizontally on the ground of our forests, the offspring mostly stay in the U-shaped living-tube or in a specially woven nursery. If a beetle is caught in the web, at first only the mother comes hurrying along. When salivating and sucking the blood out of her prey, when wrapping up a victim for storage, or when cleaning, she automatically makes the web quiver with slight, gentle, low-frequency vibrations which are characteristic. These are interpreted as call signals by the young, and make them come to take part in the feast.

In contrast to the different call signals the spider mother has a standardized warning signal [writes Dr. Tretzel]. While the movements with a call effect make the web vibrate slightly and softly, warning is given by a short, violent jolt to the web caused by rapid movement of a hind leg. This warning serves to send forward youngsters, who want to accompany their mother in her assault on the prey, back to the safe hiding-place. The mother stops briefly, moves one hind leg, and it is amusing to watch how the obedient youngster immediately turns tail and hurries back. The young are also sent back as long as a captive prey is still struggling hard.

So the spider mother's behaviour *looks* surprisingly sensible. We get the impression she is warning her young of an impending danger. But on closer study of this behaviour the evidence is all against her being conscious of a danger to her young. For her warning is omitted in other dangerous situations which can be experimentally produced; that is, when mother and youngsters are feeding on a prey on the spread-out web, and a second prey is put on the web alive, not far from them.

The mother attacks it, but without first sending the young back into the shelter of the living-tube. In this case it is enough for her that owing to the violent shocks to the web caused by the second prey, her young are sitting still. She does not give a warning, unless one of them should take it into his head to accompany her. All observations point to the conclusion that the warning is given only if the mother is hindered in her location of the prey by the young rushing after her.

Birds too can interpret certain shocks correctly. When at night a robin sits in the branches of a tree asleep, no matter how hard the wind may shake it, the bird does not take much notice. But if it feels the very slight, typical vibrations caused by a climbing marten, it starts up, trying to escape, often successfully.

Bees in their hive possess a regular system of vibration signals. We often hear a buzzing near a beehive. The bees cannot register this air sound. But with the soles of their feet they feel the vibrations of the honeycomb caused by the whirring of the wings of other bees of the same hive walking about.

Tones of body sound are possible, loud or soft, high or low. The well-known duet of hooting and croaking, between the old queen and the young one lying in the queen cell shortly before being hatched, has been investigated in more detail by an American professor, Adrian M. Wenner.[5]

In case of danger there is an alarm tone which puts all bees in an aggressive mood and stinging frenzy. If intruders turn up at the flight hole – wasps, for instance – the guards jerk forward on their legs, repeating a short sound signal every two or three seconds for about ten minutes. If one shakes the whole hive, the same signal is sounded in unison by hundreds of bees, and it is advisable to clear out in haste. As soon as the disturbance is over, squeaking pervades the hive for several minutes: all clear! Then everything is quiet again. This vibration signal, which puts the bees in a peaceful mood and stops them from going on aggressive excursions, is artificially produced by American bee-keepers with an electric buzzer attached to the hive. They can then work in the hive without being stung.

A strange "chorus," as Professor Wenner calls it, is sounded before swarming: the "prepare-to-swarm" tone. In 1959 an American engineer developed a warning device built into the hive, which reacts to these vibrations, so that the bee-keeper is informed about the impending event and can quickly catch the swarm.

By means of vibrations returning worker bees can tell those that stayed at home something about the quality of the field in bloom they have visited. There is a tail-wagging dance by which a successful worker gives information to the ones left behind on the direction and distance of the feeding-place; during this the dancers emit various rattling noises from their wings, as was discovered by Dr. Harald Esch,[6,7] of Munich University. The better, richer, and nearer the source of nectar or pollen, the faster the rattling of the bees. A change in the quality of the food gathered is indicated by a change in the number of separate vibrations if the feeding-place is near; if it is far, by a change in the duration of the intervals. An improvement, for instance, in the concentration of the sugar solution from 0·5 to 2 moles on a feeding-place a thousand yards away causes the interval to be shortened from 50 to 33 milliseconds.

The bee thereby puts distance and quality into a very sensible economical relation. Superior feeding-places at great distance are given the same value as less prolific but nearer ones. Thus the unemployed bees in the hive can tell merely by the vibrations of the nectar-advertising workers which of them is making the best offer and from which dancer it is wisest to collect more detailed information about the flight direction and distance of the promising objective.

Man, too, has a vibration sense: a better one than is commonly believed, at least according to Professor John Linvill.[8] Since 1964 he has been trying at Stanford University in California to develop an instrument for his blind daughter, which transforms light into vibrations so that ordinary print can be felt with it.

This reading apparatus for the blind consists of a little mosaic of photo-cells, which is passed over one letter after another in the printed text. All photo-cells just above a black spot cause piezo-electric crystals to vibrate. The crystals are arranged in an identical magnified mosaic and are touched with a fingertip. After some practice the twelve-year-old girl was able to recognize the letters by their vibration pattern and to read about twenty words a minute.

The Tactile Sense

A disaster nearly happened to a beehive. The busy builders had just finished a new honeycomb wall, when Professor R. Darchen[9] tilted the nesting-box on its side. The older walls remained rigid, but the new one, which was still soft, bent slowly and threatened to block the entrance to the breeding-chambers of the comb below it. This would have spelt death to the brood growing up in it.

But the bees were immediately at hand with constructive measures to prop up the bending wall. Very soon hundreds of them had formed a chain of living poles where the bend was worst. With their legs stretched out and pointing forward or back they stemmed themselves against the pressure, keeping apart the (for them) quite heavy masses. At the same time, other building bees started making pillars from wax, which when finished could completely take over the task of the living poles.

It is difficult not to regard these repair works as actions governed by reason. Professor Rémy Chauvin[10] makes the following comment:

> Darchen's papers are of particular interest because they have shown us that some traditional views about animals being subject only to their instinct do not agree with the facts. The idea that bees work like machines regulated in advance once and for all is thus shown to be completely wrong. In reality it is all much more complicated. We are dealing with a social group. which recognizes difficulties and can solve the problems facing it. So here is no mechanical process dependent on the instinct, but an activity of a higher order. In this context, however, I would not like to talk of intelligence either, because for that phenomenon a number of other things are needed as well.

But it is established that bees have an absolute feeling for some architectural measurements, including the distance that two adjoining cells must have from each other. As soon as this distance changes, the bees try to restore the normal state by every means available, including the miner's method of propping up the ceiling.

The bees also have a reliable feeling for the size of the hexagonal cross section of the honeycomb cells. Cells designed for a drone brood are larger than the ones the worker bees are to grow up in. So every time, before she lays her eggs, the queen measures with her tactile abdominal bristles to discover to which kind the empty cell belongs. If it is small, she opens a valve-like closure in her abdomen shortly before the egg slides out, and fertilizes it. If she finds the cell is larger, she does not fertilize the egg; which means it will turn into a drone. The working of this mechanism is exact to the tenth of a millimetre.

Dr. D. Merrill[11] has made interesting experiments at the University of Michigan with similar hairs for measuring lengths on the abdomen of caddis-worms. These larvae live in running inland waters and build nests and a very ingenious case, which they carry with them as a snail does its shell. By sensory hairs at the tip of their tails they feel if the tube is the right size. After Dr. Merrill had cut these little hairs off the larvae, they went on and on building, like a woman watching an absorbing programme on television, who knits a sleeve much too long. They didn't stop until the silken structure, interspersed with stones and leaves, was three times as long as necessary, and was probably too heavy for them to go on.

The ichneumon-fly *Pimpla contemplator* also finds out its victim's measurements before injecting it with an egg. The ichneumon-flies hatching from large meal-worm pupae are always female, while the males come from the smaller ones.

For a long time experts were baffled by the infallible way the hunting wasp *Philanthus triangulus* always found the weak spot in its prey. This predatory digger wasp attacks bees in flight, throws them to the ground, paralyses them by a quick sting, then flies off to its nest with them.

The strange thing is that the hunting wasp's sting is absolutely unable to penetrate the bee's armour. But on the bee's abdomen, immediately behind the first pair of legs, there is a tiny soft-skinned place, which is hardly visible; the predator has to be as quick as lightning in finding and hitting it. In 1962 D. W. Rathmayer[12] discovered the secret. Directly at the sheath of its sting the hunting wasp has a highly specialized tactile organ designed solely to locate this unprotected spot. As soon as it has spotted that, it stings.

These examples alone give some idea of the many different ways in which the tactile sense can be applied. With a primary function, no doubt, of registering collisions and contacts, it has developed into being also a sense for lengths and shape in recognizing forms both specific and general.

Many water-dwellers also have an extended tactile sense for aggressive purposes. The jelly-fish called Portuguese man-of-war certainly lives up to its martial name with its fishing lines a hundred feet long. As soon as a fish up to the size of a mackerel has got caught in them, the man-of-war fires its nettle batteries with a numbing effect and draws its victim up to its feeding polyps. But if it feels a fish of the genus *Nomeus* at its fishing threads, it refrains from all hostile action.

The huge antennae of deep-sea fish, with which they can locate a swimming prey, are also like nerve-fibre spiders' webs. Other long-distance feelers are the antennae of crab and shrimp and the barbel of catfish, enabling them to take their bearings by touch, which they do astonishingly well. Cats, too, can recognize in the dark any object they touch with their whiskers, just like a man who feels them with his fingertips. Only in cats this sense is more finely developed and works faster. For instance, according to Lorus and Margery Milne[13], if the cat's whiskers touch a mouse, the cat reacts with the speed and precision of a mousetrap.

With the desert jerboa its overlong whiskers play the part of a blind landing device. The animal is active at night and with its two enormous hind legs can jump like a kangaroo, so fast that in the headlights of a car it looks like an arrow flying over the ground. It can easily stumble in the dark if the ground is bumpy; the Viennese zoologist, Professor Otto Koenig, describes how it guards against that:[14]

> During the jump the animal points two whiskers nearly as long as its body straight downwards and so always remains in tactile communication with the ground. In this way it can prepare its legs for any hollow, any stone or bush, and if necessary can suddenly change course while still in the air by using as a "vertical rudder" its long tail, which is also in permanent contact with the earth.

Direction and Wind-Speed Sense

Now and then scientists investigating animal anatomy discover curious sense organs whose purpose they cannot at once detect. Such was the case in 1957 at the Zoological Institute of Würzburg University, where two neuro-physiologists, Dr. Dietrich Burkhardt and Dr. Günter Schneider,[15, 16] were working on that commonplace creature, the bluebottle.

It was known that in many insects a small number of sensory nerve cells are to be found in the two joints that connect the antennae with the head; their purpose, however, was not known. To discover it, the two scientists carried out an experiment.

They bound up a bluebottle with a special "belt" and took it like an aircraft into a small wind-tunnel. Beforehand they had inserted microscopically small electrodes into the single sensory nerve cells in the antenna joint, to intercept the nerve signals sent from there to the insect's brain.

At wind force nil, complete silence prevailed in its antenna nerves. But as soon as a light wind passed gently over it, the sensory nerve cells began to transmit impulses at regular intervals. With increasing air speed there was a steady increase in the speed of transmission.

So the bluebottle has in its antenna joints (the so-called Johnston organ) a highly efficient wind-force indicator. The air stream presses the two antennae backwards against the muscular tension. This tilt is noted by the sensory cells in the antennae joints and converted into a corresponding series of nerve-signal volleys.

The next experiment confirmed this finding. On a windless day the scientists pressed back the two antennae with a pair of tweezers. The sensory cells immediately telegraphed exactly the same signals to the bluebottle's brain as if the antennae's slanting position had been caused by the wind. The insect's reactions too were the same: it placed its wings in the flight attitude – one might say retracted its undercarriage – and, according to how far back the antennae were pressed, it changed the figure of eight described by its wing-tips during pulsation and the starting angle of its wings. Both are measures taken by flies to change their flying speed or buoyancy.

To discover how this meter operates in practice, the scientists pressed the antennae back and fastened them firmly in that position; then in the laboratory they made the insect fly over a "race track" towards the window. They established that the farther back the bluebottle's antennae were bent, the more slowly it flew. The electric nerve signals were thus reporting to its brain a wrong flying speed. It thought it was flying too fast and adjusted its wingbeat in such a way that its flight was slowed down. All this confirmed the hypothesis that the bluebottle uses its wind-speed indicator to determine its speed of flight – a conclusion not so obvious as it may at first appear, for there are many other insects that have a different use for it.

For some insects it is also important to get information about the wind-speed from a "private meteorological station." The blow-fly (*Lucilia sericata*), a relative of the bluebottle, before taking off, feels with its antennae whether the wind is blowing faster than 100 inches a second; in which case it will not fly up at all. Since it always has to fly against the wind in making for scent sources, such a flight, when the wind is faster than the insect can go, would only take it farther away from its goal. Butterflies behave similarly (see p. 132).

The dung-beetle, which is socially quite acceptable in zoological circles, with its long legs does not "hold the road" at all well, and is markedly sensitive to side winds. Consequently, it finds a wind-force and wind-direction indicator very useful. This was discovered by Professor Georg Birukow[17] in a series of interesting experiments at the Zoological Institute of Tübingen University. If the wind is directed over the beetle from the front, it bows its head forward; if the wind come from behind, it raises its head high. In a cross-wind it tilts around its longitudinal axis and with its smooth back leans against the wind as if it had a "list." The greater the wind force, the more pronounced and vigorous are the attitudes it adopts to offer minimum resistance to the air current and improve the pull of its legs.

Things become more complicated with the honey-bee, if a cross-wind threatens to drive it off course when it is making a beeline for a known source of nectar. As Professor Herbert Heran[18] of Graz University established in 1964, the Johnston organ measures the bee's flying speed in relation to the surrounding air, while the facet eyes, together with special nerve circuits, determine speed over the ground by parallel earth soundings. From both speed values the insect instinctively works out the angle it has to hold against the wind so as not to be driven off course. (See also p. 53).

Moreover, the bee still remembers this holding angle when returning to the hive, computes it with the angle of its course to the sun, and thereby works out the right direction of flight, not for the wind conditions that prevailed on the outward trip but for calm conditions. Since wind-speed and direction may change at short notice, any unemployed bee has to learn from the successful homing forager the exact, undistorted beeline, and then re-calculate for itself the angle against the wind it must fly – always depending on the wind conditions then prevailing.

The process seems to be complicated a good deal by the fact that the wind is often gusty. But in 1965 a discovery by Volker Neese,[19] a pupil of Professor Martin Lindauer at Frankfurt University, brought a remarkable solution. Under a strong magnifying glass one can see a number of fine hairs sprouting out of the facet eye. These are sensory hairs, which can be bent like levers by any gust of wind. They enable the bee to measure any incidence of gusts and their force, so as to balance them by flight manoeuvres carried out at lightning speed.

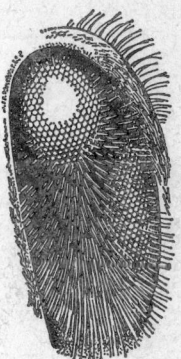

Figure 78. The bee's eye has a hairy coat! Between the 2,500 ommatidia of the compound eye many fine sensory hairs stick out into the air. They are so arranged that they do not block the bee's view, and they measure each gust of wind so that the insect can react with lightning-swift flight manoeuvres to avoid being driven off course. The hexagonal pattern of the compound eye is visible here only at two places where the light is reflected.

Chapter Seven

Gravity, Hunger, and the
Electric Sense

The Sense of Gravity and Balance

AMERICAN psychologists can gather information about a person's character by the way he straightens a picture which is hanging crooked. This is connected with two different senses by which we judge whether we are standing upright and whether or not other objects are in a perpendicular position. One is an internal sense of gravity and equilibrium, the other is an optical comparison with our image of the environment. In varying degrees everybody prefers one or the other way of taking his bearings. It happens unconsciously and reflects essential features of the personality. As with the optical sense (see p. 17) and the sensation of pain (see p. 101) we find that sense impressions are partly governed by character.

Yet it seems at first as if the perception of gravity is a physically exact measuring process which could not be influenced by anything.

For this process nature has prepared two types of construction which are different in kind. Variations of the first are to be found in man, mammals, birds, and fish, as well as in worms and primitive molluscs. The second type is reserved for the insects.

With bivalves this sense organ is constructed with a most

ingenious simplicity. These molluscs too have to know exactly where up and down is when they dig themselves into the sand, and more particularly when they emerge from it again. The information is provided by a little hollow sphere which is padded inside with fine sensory hair cells. A small round stone can roll slowly round in the cavity, which is filled with a braking fluid. By gravity, that is, by the attraction of the earth, the stone always rolls to the lowest point of this sensory vesicle, which is called a statocyst, and stimulates the sensory hair cells there.

Apparently some of the sensory hair cells, when stimulated, tell the clam: "Now your body is in its proper position. Stay like that." All the other hair cells when stimulated immediately trigger off movements which shift the clam as fast as possible back into its right position.

Till 1958 scientists had been looking in vain for such or similar sense organs in insects. There was no doubt, however, that insects could register gravity. As Dr. Detlef Bückmann[1]

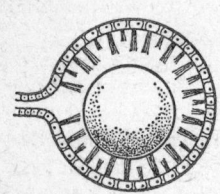

*Figure 79. The organ of balance of the St. James's shell (*Pecten jacobaeus*), the so-called statocyst, is a device of ingenious simplicity. If a small round stone, in the cavity padded with sensory hair cells, touches particular hair cells, their impulses are taken by the nervous system as information that the shell's body is in a state of balance. Impulses from all the other sensory hairs always aim at making the shell return into that state as soon as possible.*

put it as late as 1956, it was a question of finding the still unknown sense organ underlying a known function.

The scientists' quest took so long because they had paid too little attention to one feature of the insects' body; the fine hair in a few places. But in 1958 Dr. U. Bässler[2] found the same answer for mosquitoes and meal-beetles as did Professor Martin Lindauer and Dr. J. O. Nedel[3] for the bee a year later, and Professor Lindauer's assistant Dr. Hubert Markl[4] for the red ant in 1962: all hair cushions on the joints of the limbs consist of sensory hairs. In principle they serve the same pur-

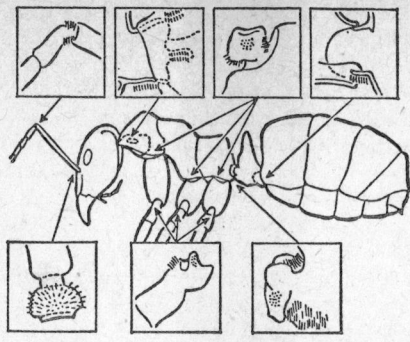

Figure 80. The ant has fine pads of sensory hairs on all its joints, as can be seen on the section enlargements. The more a part of the body is deflected from the normal position, the more hair it stimulates. Through nerve fibres they signal to the brain the position of each limb.

pose as the sensory hairs inside the statocyst, while the part of the plumb bob that stimulates these hairs is taken over by the limbs.

Strictly speaking, this is an indicator of limb positions, informing the insect first of all about the positions of its head, antennae, legs, and abdomen. But whenever several limbs are deviated in the same way, against the normal muscular tension – that is, drawn downward by gravity – the ant's brain works out from that the gravity direction. Without this information it could hardly find its way about in the maze of the ant-hill.

Figure 81 represents a comparatively simple example. A deflection of the abdomen by 20 degrees to the side indicates to the insect that the axis of its body is at an angle of 45 degrees to the perpendicular. But as mentioned above, the information must first be confirmed by other joint-position indicators. Otherwise the ant will interpret the message differently, for example, as the influence of a heavy burden which it is carrying.

How does all this work in man? Our balance organ is closely connected to the inner ear. Figure 82 shows a diagram of all its essential parts.

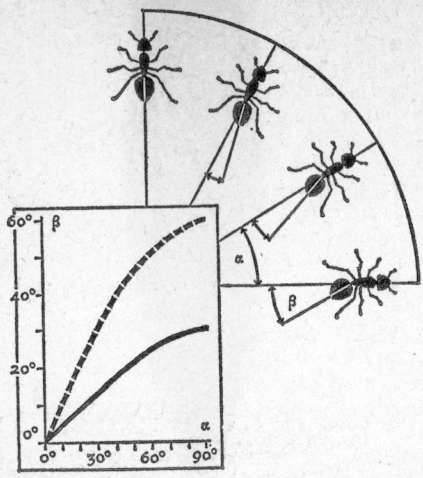

Figure 81. The deflection of the abdomen by gravity is used by the ant's brain as a protractor for the angle of the body to the perpendicular. From a deflection of the abdomen by an angle (β), the ant can always recognize that it deviates from the perpendicular by an angle, α (continuous graph). If the sensory hairs are removed, the deviation is twice as great (dotted line graph).

In each of the two vestibular sacs, the *utricle* and the *saccule*, there is a field of fine sensory hair cells. Just as a wheat stem bears a heavy ear at its top, the sensory hair cells bear tiny calcareous crystals. Under the influence of gravity these bend the hair cells, stimulating them in a characteristic way. In principle this is a refined version of the clam statocyst and represents the balance organ proper.

Quite a different purpose is served by the three semicircular canals. They are also filled with the inner ear fluid, the endolymph. At its entrance each canal has a row of sensory hair cells – which are not weighed down by crystals. If a man turns or shifts his head, something happens in the canals that can be observed in any cup of tea: the fluid does not participate in the movement. But since the canals shift, the sensory hair cells register a stimulus.

Figure 82. The labyrinth of the sense of balance. In the two vestibular sacs filled with fluid, the utricle (A) and the saccule (B), there are pads of sensory hair cells as the actual balance apparatus. The rows of sensory hair cells at the entrance to the three semicircular canals (C) register every movement of the head. The nerve tracts (D) lead directly to the eye's nervous system and to the brain. (E) shows hammer, anvil, and stirrup again, (F) shows the cochlea of the inner ear.

So exact three-dimensional geometry is at work here. All three semicircular canals are exactly at right angles to each other, thus dividing any movement of the head into three parts: the horizontal one, the upright perpendicular one, and the transverse perpendicular one.

What is the purpose of this complicated mechanism? Dr. Hellmuth Decher,[5] lecturer and head physician of the University Clinic in Bonn, gives a very impressive description of it:

There are innumerable nerve connections leading from the balance organ and the semicircular canals to the cerebrum, cerebellum, brain stem, and spinal cord, as well as to the eye muscles and probably to other regions of the nervous system as well. It is confusing how many sensory and nervous functions are here closely interconnected.

We may be conscious of the sense for above and below; but it also works unconsciously at the same time. As soon as a standing person threatens to fall out of his position (which is unstable from a static point of view) – and that is happening all the time – the balance organ immediately mobilizes muscle groups in the legs, trunk, and arms to counteract the fall.

The control system, however, must also be capable of distinguishing between a mere nodding of the head and a tumbling movement of the whole person. This is what the semicircular canals are for.

You may have felt rather incredulous to read on p. 171 how even when the head is shaken the optical impression of a fixed image is preserved. I mentioned a mysterious calculating mechanism which compares the messages of the visual nerves with the head's statements of position. Well, in the sensory nerves of the semicircular canals and in their co-operation with the optic nerves, we have come upon this calculating mechanism.

For the time being we can only suppose that the sensory nerves of the semicircular canals are also connected with the auditory nerves. In an open environment without an echo we can recognize immediately whether a noise is coming from right or left – I have discussed direction hearing in detail. But we are completely unable to decide whether a noise is coming from the front or back unless we move our heads to and fro a little while we are listening. So it is at least conceivable that changes in loudness at different positions of the head are weighed against each other, thereby producing the impression "in front" or "behind."

If with certain diseases the nerves of balance and of the semicircular canals are incapacitated, we are overcome by vertigo. We fall down; pictures swing before our eyes, and we can no longer hear which direction a voice is calling from. Nothing could demonstrate the value of a sense better than the failure of just this capacity.

So far everything still looks like exact physics and "nerve mathematics." Until the results of his experiments made him think again, this was also the opinion of Professor Herman A. Witkin[6] of the Downstate Medical College of the State University of New York.

He rightly told himself that, besides the sense of balance, the eye too is constantly watching over our position in space. It takes its bearings from the corners of a room, from trees, lines it is used to, the horizon, and any other possible clues. Admittedly the sense of balance can manage without the eye, for example, when we drop our lids. But as soon as we open our eyes, they assist the internal sense.

How large is the contribution made by these two senses to the general impression of uprightness? To find this out, Professor Witkin created a highly unnatural world in his

Figure 83. Experimental apparatus for confusing the human sense of balance. The test subject is strapped to a chair which can be tilted sideways, and is in the middle of a room which can also be tilted sideways independently of the chair. In these experiments many people lose their sense of the upright.

laboratory, one in which the two senses were no longer complementary but worked against each other.

By the apparatus shown in Figure 83, he distorted the eyes' influence. The subject of the experiment was strapped to a chair which could be tipped any distance to the left or right. All the subject could see was the inside of a little room with the fourth wall missing. This room could also be tipped sideways (independently of the chair). After both had been tilted in the dark, the light went on and the experimenter now kept adjusting the chair till the subject felt he was sitting absolutely upright.

This was, in fact, very rarely the case. Some subjects were then suspended at an angle of 35 degrees. In an extreme case, the person even had a list of 52 degrees when he replied with a confident "Yes" to the question, "Is this the position in which you eat at table?"

In these cases the test subjects had unconciously relied more on their eyes than their internal sense of balance. But the moment they closed their eyes, they suddenly felt aslant, and now, following only their sense of gravity, they manoeuvred themselves with great exactitude into a position that was really upright.

A second series of tests was then carried out to disclose how the same test subjects judged the uprightness of other objects – by their own sense of gravity, or by the image of their environment. Again they had to sit on the tilting chair, and in the completely dark laboratory there were only two things left for them to see: a frame covered with fluorescent paint and inside it a fluorescent rod. This was put into an exactly upright position.

Again it proved that a number of the subjects more or less strongly preferred either the environment – that is, the position of the frame – or their own sense of gravity. But what puzzled Professor Witkin was the discovery that in spite of this great difference of reactions from one person to the next, every individual produced completely identical results in both types of experiment – identical, that is, to a single degree of the angle! Anyone who in the first test had taken 32 per cent of his bearings from the optical impression of his environment, and 68 per cent from his internal sense of gravity, did this in the same proportion in the second test as well.

It is amazing that man registers the position of his own body in the same manner as the position of foreign objects, for example, a picture hanging on the wall. By taking measurements it would be possible here to work out a "personality constant" expressed in percentage rates. It was of course tempting for a psychologist to guess at an underlying psychological process, by which we form the impressions about ourselves and our environment.

It is impossible to describe here all the sequences of psychological tests whereby Professor Witkin systematically confirmed his hunch through a process of elimination. His findings were that there are indeed direct relations between the way a man registers uprightness and his personal qualities.[7]

On might divide all humanity into those who are "independent of the field of vision," others who are "dependent on the field of vision," and a wide spectrum between these two extremes. People independent of the field of vision excel at solving difficult problems, since it is comparatively easy for them to pick out the essentials from complicated contexts and combine these with other elements into a new shape. People absolutely dependent on the field of vision, like the subjects with a heavy list who still believed they were sitting upright, possess very little of this analytical and creative faculty.

Furthermore, a social aspect became apparent: those who are independent of their field of vision are also rather independent of their environment and of other people. They cannot very easily be made to conform to social conventions, they do not follow every fashionable trend, but in extreme cases are bad mixers, asocial, or eccentrics. Conventional people, on the other hand, always take their bearings from their environment, in the purely physical sense also. Women on the average are more inclined towards field dependence than men.

Small children are almost completely dependent on their field of vision. The development towards independence starts at about the eighth year, then accelerates, according to the child's talents and education, to reach a temporary climax by the thirteenth year. This level is maintained till the age of seventeen, when something unexpected happens. With most adolescents the trend reverses. The outer compulsion to fit in, to subordinate themselves, to conform in their jobs and in society, has its psychological effect. Comparatively few adolescents go on developing their tendency towards individualism.

Strange that such things can be traced merely through noting the senses by which people register the upright.

The Senses of Hunger and Satiety

Many overweight people are on the horns of a terrible dilemma. Either they eat far too much – openly or secretly – or they lose weight at the high price of constant pangs of hunger.

It would be comforting for them to blame their trials and tribulations on some gland deficiency or fault in the metabolism; and of course this is sometimes the case, which makes the condition easier to cure. But unfortunately scientists have not so far found any single metabolic process that might be considered a general cause of obesity.

Perhaps the root of the trouble is not in the body's food-processing machinery at all, but in the gauge of fuel and substances required: in a wrong assessment of conditions, that is, by the sense of hunger and satiety.

What is it, anyhow, this dull feeling of hunger, and how does it originate? Before the First World War this seemed an easy question to answer. The Americans Dr. W. B. Cannon and Dr. A. L. Washburn[8] inserted a balloon into a dog's stomach and kept blowing it up and letting it shrink again. When the balloon was inflated, the dog always behaved as if completely satisfied; even if it had not eaten anything for days, it would not touch a dish of food. A few seconds later, directly the air had been made to escape, it would attack the food ravenously.

So the feeling of satiety must be connected with the extent to which the stomach has expanded. The musculature of the stomach wall is in fact streaked with sensory nerves which respond to expansion and signal the state of tension to the brain – and the same applies, incidentally, to nearly all other muscles.

But there is more to it than that. Professor C. Kayser[9] of Strassburg describes an interesting experiment in which a dog was fed not through its mouth, but by a plastic tube inserted from outside direct into its stomach. The animal needed almost twice as much food as usual before it gave signs of satiety. So besides stomach expansion, Professor Kayser concluded, the sliding of food through mouth, throat, and oesophagus also plays its part in the feeling of being satiated.

But there is still something else. After all, everybody knows that bacon and eggs fill up the stomach much more than an equal volume of water. The body, therefore, must possess yet another gauge by which to determine the nutritional value of the food it has taken in.

How does this work? Brain surgeons found on the lower side of the brain, in the centre of the hypothalamus, a region that may be called the satiety centre. It is in the immediate vicinity of the "thermostat" controlling the body temperature (see p. 89) and the thirst centre which registers the salt content of the blood plasma and accordingly controls the excretion of water and the urge to drink.

Scientists have destroyed this centre of satiety in rats, mice, cats, dogs, and monkeys. The results were always the same: the animals developed a terrifying voracity. During the first month after their operation they doubled their weight. Everything has its limits, however, and after growing so enormously fat, they stopped putting on weight. Slender comfort for obese people!

Two facts emerge. First, the satiety centre is by its nature an anti-overeating mechanism. Secondly, we have an assurance that if this inhibitory mechanism should ever fail, the body still has a sort of emergency braking system, obviously in the shape of further satiety centres which are informed about the stocks of fat reserves.

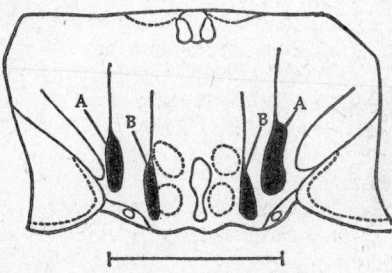

Figure 84. The hunger centres (A) and satiation centres (B) are shown in this frontal cut through the centre of a rat's hypothalamus.

Is there a hunger centre as well? Yes. That too is situated in the hypothalamus, about half a millimetre from the satiety centre. If this hunger alarm is destroyed, the reverse happens: the experimental animals refuse any food and soon die of starvation, without feeling hungry. Rats lose 45 per cent of their original weight after nine days. So the decline comes even quicker than obesity.

How is it possible at all that the centres can be informed about the body's nutritional condition? At present there is much evidence in favour of Professor J. Mayer's[10] theory, which is roughly as follows:

Every nerve cell is by its nature already a sort of gauge for the sugar content of the blood. In contrast to muscle cells, they have no store of sugar, although they need it constantly to be able to function. Therefore they must take this fuel continually from the blood. If the blood has too little sugar, the nerves fulfil their tasks imperfectly. Everyone who has ever been really hungry for any length of time will be able to confirm that this state also has a paralysing effect on thinking. But a small amount of grape sugar immediately revives it.

Specially sensitive reactions are shown by the nerves of the satiety centre, which has thus become a sugar gauge. These special nerves develop an immense "craving for sugar" and take considerably more sugar from the blood stream than the nerves of the hunger centre. It has already been proved in live animals that a large content of sugar increases the rate of firing of the nerves in the satiety centre and inhibits the electrical activity in the hunger centre.

To be precise, these two complementary effects are not a measure of the absolute blood-sugar content – for in this case diabetics ought never to feel hungry – but of the difference between the sugar levels of the arterial and the venous blood. It is very complicated, no doubt, yet any simpler solution would endanger our existence.

But there is even more. Our sense for hunger and satiety comprises a forecasting department. In hot weather men and animals naturally need less fuel for keeping the body warm than in cold weather. Yet the messenger services described earlier would inform the satiety and hunger centres about the changed requirements only after some delay, perhaps even at a time when temperature conditions had changed once more.

So a direct news service from the thermostat (which is right by them) provides temperature forecasts informing these centres of incipient changes in the sugar level. Apparently, the way the thermostat changes affects the regulation of the appetite, the exact amount of food which can be saved by a rise in the internal temperature will automatically be felt in advance.

Scientists have checked this by experiments with a goat. They were able alternately to increase and reduce the animal's voracity at will, first cooling the thermostat by the insertion of a water-pipe and then warming it.

One American scientist who shall be nameless applies these

findings to unwelcome guests he has been obliged to invite: he receives them in overheated rooms, because then they do not eat so much at meal times!

Even at this stage we have still not completely exhausted the phenomenon of hunger. For in the regulation centres nerves converge also from other regions of the brain. Some of them, for instance, come from those parts of the brain responsible for the behaviour connected with self-preservation or the preservation of the species. Among other things produced by this convergence is the wide-spread phenomenon of competition for food.

Moreover, hunger and satiety are connected with feelings of pleasure and displeasure which have their seat in the region of the brain called the limbic system. So in this field too the channels have been traced by which psychological processes affect physical ones.

Many obese people do not like this idea at all; for (as doctors are well aware) overweight patients would rather blame their fat on organic than on psychological reasons. From all I have said in this section, however, it is clear that the regulation of the appetite may be disturbed in many places; so at least in theory there may be many different reasons for obesity, including psychological ones.

Professor Arthur Jores[11] and his assistants, Adolf Ernst Meyer, Herbert Maisch, and Dr. Freyberger, all of the Hamburg University Clinic, have made a thorough study of this latter aspect. They find that obese people in their childhood had often received food or sweets from their parents to help them over emotional upsets. In this way food gradually became a means of assuaging feelings of *malaise*.

In an experiment these Hamburg doctors made forty patients fast, who were between 50 and 90 per cent overweight. After a short time half of them complained of depressions and anxiety feelings, which promptly disappeared as soon as they started eating again. With at least half the test subjects, the absorption of food obviously served to ward off feelings of *malaise*.

This conclusion, unfortunately, is not yet much use in practice. Attempts to cure obesity on the psychotherapist's couch have been successful in discouragingly few and exceptional cases; according to the Hamburg doctors, the defective behaviour implanted in childhood is too deeply rooted.

How Many Senses Are There?

How many senses are there? The reader may find it amusing to look through this book and add them all up. But please do not forget the senses that have been mentioned here only in passing, such as the sense of thirst in man and the sense of carbon dioxide in bees.

Anyone who tries to count the senses will soon encounter great difficulties. The camera eye of man, the camera eye of the frog, the facet eye of the bee, and the eyespot of the bee – are we to regard these as one, two, three, or four senses? Or even five senses, since two entirely different types of nerves, rods and cones, are necessary for seeing light and for seeing colours.

Perhaps our way of hearing is a combination of several senses, that is, for loudness, direction, pitch, and for under-standing language and music. The measuring of stomach expansion, the sensation of the food gliding down the oeso-phagus, the measuring of nutritive value and of the blood-sugar level – are these four or two senses, or perhaps only one?

Are we to base our classification and count on the way the sensory cells are stimulated, and to list under the heading "mechanical senses" touch, hearing, the sense of gravity, and the sense of pain, or should smell and taste also be included? This would be too summary a treatment to have any value at all. Or should we take the "attachments" of the sensory cells as a criterion – that is, all those little hairs, membranes, pores, fluid-filled tubes, labyrinths, spirals, and so on? In nature their types of structure are legion. Should we regard the way the stimulus is processed by the brain as the essential feature? This would be an even more impossible task.

Besides, there are a great many senses I have not mentioned at all, such as the bee's atmospheric humidity meter,[12] the fly's gyroscope, the pressure gauge in the kidney.[13] This book would be at least twice as thick if it were to discuss all the body's internal senses: the mechanism for narrowing and widening the large blood vessels and for closing and opening the capillaries,[14] the lymph regulators, the pacemaker for the heartbeat, the sense of choking and the drive to breathe,[15,16] the activation and inhibition of innumerable hormone secretions, the indicators for the muscles' state of tension,[17] the recognition of foreign bodies and the control of antibody production[18] – and so on for several pages.

Within the framework of this book, I must confine myself to only a few of these; and among senses which receive their stimuli direct from an external object, I shall deal with only one, which is most unusual, the electrical sense.

Besides the direct method of taking bearings, there is another, completely different method in the animal world, so-called "signpost orientation." To us human beings it has always seemed particularly enviable. When storks or warblers are flying from Europe to distant Africa, they can neither see nor hear nor smell nor feel nor in any other way directly perceive the goal of their flight. They have to adjust their course according to other clues and signposts. We shall take this matter up after covering the electrical sense.

The Electric Sense

Can fish be caught with a magnet? I do not mean cardboard ones with metal mouths like the ones in children's game sets, but real live fish. It sounds highly improbable, and yet it has been done – first by the Cambridge zoologist Dr. H. W. Lissmann.[19]

Drifting in a rowing-boat on an African inland waterway, he discovered a knife-fish, *Gymnarchus niloticus*, a magnificent specimen more than five feet long. When he was within a foot and a half of it, he held a strong horseshoe magnet just above the water level. As if attracted by the magnet, the big predator fish approached and stayed put with its head directly under the magnet. When he moved the magnet to and fro, the fish would follow slowly. In the same way, though less strongly, it reacted to the movements of an hard-rubber comb, which the professor had electrified by combing his hair.

This amazing trick could not be carried out with a pike or most other fishes, only with the African knife-fish, the elephant-fishes of Africa and the knife-fishes of South America. What is the explanation?

Gymnarchus niloticus belongs to a small group of fishes, which have very limited vision, whose hearing is poor, and which do not rely either on the pressure-wave receivers of their lateral-line organs. They take their bearings in a most peculiar way, by electrifying their environment: they find out what is going on around them by the way that environment distorts lines of the electric field from the course they would take if undisturbed. This is probably the sensory world which for us humans is the strangest and most difficult to grasp. It is so unfamiliar that even in measuring techniques there is nothing really comparable. It has nothing at all in common with radar or sonar, and is also fundamentally different from the ultrasonic world of bats and dolphins. The electric fish does not receive echoes nor does it measure time lags.

Instead it possesses a sense that we completely lack. At school the science teacher always has difficulties in making clear the concept of the electric field. A magnetic field can, after all, be made visible easily by iron filings. Demonstrating the electric field by similar means is better omitted because dangerously high tensions of several thousand volts may be needed.

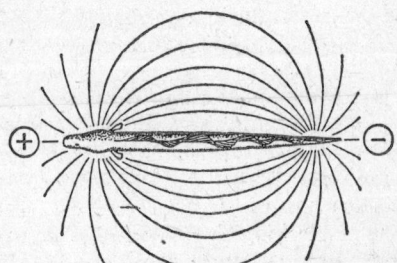

Figure 85. The electric lines of force of Gymnarchus niloticus *run like the lines of force of a bar magnet, unless diverted by foreign bodies.*

Thus the electric lines of force remain to us a somewhat uncanny abstract idea. We cannot doubt their existence, but they cannot be seen, heard, tasted, felt, or experienced in any other way. For *Gymnarchus niloticus*, however, the lines of the electric field are the principal medium of information, like light for us.

Yet the fish does not work with high voltages like a science teacher; not even with 500 to 800 volts like the electric eel, 450 volts like the electric catfish, or 50 volts like the electric ray. It produces in its "living batteries," depending on body size, only between 3 and 10 volts pulsed direct current. The positive pole is in its head, the negative one at the tip of its tail.

The pulse rate, however, is terrific. The electric organ is continuously emitting 300 impulses a second through nearly all the creature's life, day and night, in sleeping and waking. It is like a heartbeat, only much faster.

Yet this continuity of emission has one unpleasant result. If two neighbours trespass on each other's electric field, they interfere with each other's reception just like two radio transmitters sending their programmes on the same wavelength. But in contrast to the radio engineers, *Gymnarchus niloticus* knows very well what to do about it, as Dr. Lissmann has established.[20] In the first moment of the interference the fish briefly cease transmitting. Then they abruptly change their pulse rates so that each is working at a different frequency and can therefore distinguish the other's impulses from its own.

Every impulse produces a spherical field of electric force around the fish, as indicated in Figure 85. Every object in the water, however, distorts this normal pattern of lines of force; any plant, any stone, any strange fish – or Professor Lissmann's magnet and electrified comb. All objects which are better conductors than water converge the lines of force; all poorer conductors cause them to diverge. As Figure 86 shows, this also changes the density of the electric lines of force entering and leaving *Gymnarchus*'s body.

Now this fish can register the density of these lines of force and draw the right conclusions from them about what is happening around it as far as some three to six feet away.

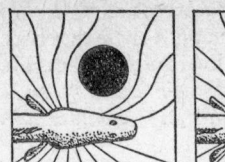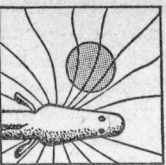

Figure 86. Gymnarchus *recognizes its environment only by the different way the electric lines of force enter its head region. A bad conductor (black ball in left picture) thrusts the lines of force apart, while a good conductor (dotted ball in right picture) draws them together.*

It can thus recognize even a glass rod two millimetres in diameter and distinguish between two objects of similar shape although they are made from different materials.

Professor Lissmann has shown the extreme sensitivity of *Gymnarchus*'s electric senses. The fish reacts to even as small a drop in voltage as 0·03 millionths of a volt per centimetre; which corresponds to a change in the current density of only 0·04 millionths of an ampere per square centimetre. This is so infinitesimal that no concrete comparison is possible. But one might say that for sensitivity the electric sense is quite the equal of an eye which can be excited by a single quantum of light; an ear which registers vibrations of subatomic magnitude; or a butterfly's antenna which reacts to a single molecule.

What makes this sensory feat so magnificent is not only the registering of electric fields, but also the correct interpretation of the field intensities received all over the body. We should remember that the patterns of these differences in field intensity are the sole stimuli in the environment by which the creature is affected.

Let us look, for instance, at the left-hand drawing of Figure 86. Nearly the same deviation of the field lines by the one object which is a bad conductor could be achieved instead by two good conductors placed at either side of the bad one. So two completely different things can produce nearly the same stimuli – nearly, but not quite. By these minimal differences the fish has to recognize a totally changed environment: an almost incredible task for a fish brain.

Yet the fish brain carries it out. By far the largest compartment in the brain of *Gymnarchus* is concerned only with the processing of electric sensory stimuli. With elephant-fish, which live in African inland waters and also locate by electric fields, the electric area, like man's cerebrum, has outgrown all the rest of the brain.

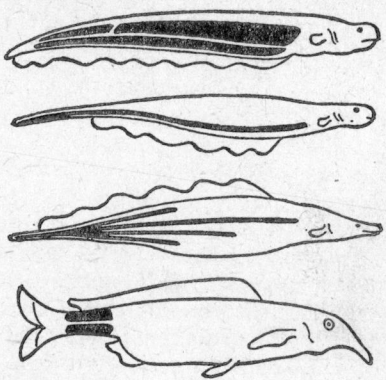

Figure 87. The 'living electricity generators' (black) in four different electric fishes. From top to bottom: the electric eel with three separate electric organs; a South American knife-fish; the African knife-fish Gymnarchus *investigated by Professor Lissman; and the elephant-nose fish.*

Elephant-fish, incidentally, are the only fish which, like the dolphin, have a pronounced urge to play. Since playing needs intelligence, the brain of the elephant-fish seems capable also of achievements of a higher order.

Gymnarchus, however, has to introduce a radical simplification of its measuring technique. If in swimming it made sinuous movements like all non-electric fish, its brain could not possibly interpret the field intensities received. So all slightly electric fish – *Gymnarchus*, the elephant-fish (about one hundred species) and the South American knife-fish (also about one hundred species) – swim as if they had swallowed a broomstick. Their spines remain rigid even while they are turning. They are propelled by wave movements of their long fins.

The most important questions, however, are for the time being still obscure: how do these fish produce the electric current,[21] and by what means do they perceive it?

Figure 88. The electric sense organ is embedded in the skin of Gymnarchus. *A microscopically small tube leads from outside into a cavity. Both are filled with a gelatinous mass which concentrates electrical lines of force like a lens. At the bottom of the cavity there are the nerve cells.*

Professor Lissmann found all over the fish skin – at the head, abdomen, and back – fine sensory pores which are arranged in fairly regular patterns almost two millimetres away from each other. Externally they are indistinguishable from taste organs, which with fish are not only in the oral cavity, but also on the outer skin. But these organs signal to the brain only if they are excited by electric fields. The pore of such an organ is the entrance to a tiny tube, which is filled with a jelly-like mass. This jelly is a particularly good conductor of the electric current and so collects the lines of force like a lens. After 0·1 millimetre the tube runs into a spherical cavity which holds the electric sensory cells.

How are the electric stimuli here transformed into electric nerve signals? Up to now scientists have found this as hard to explain as the transformation of stimuli in the visual, tactile, auditory, olfactory, and any other sensory cells.

Why did the electric fish acquire this sense, which is as extraordinary as it is complicated? Why do they not use their eyes or lateral organs like any other fish?

To all intents and purposes, the eye serves them only as an indicator of day and night, since they live in extremely turbid waters where it is very difficult to see anyhow. Also, they usually do their hunting and foraging by night. But that is still no reason why the electric sense should be necessary. Blind fish in the eternal darkness of subterranean waters can, as we know, manage very well with the pressure-wave receiver of their lateral organs.

It is another fact which is decisive: nearly all slightly electric fish live in fast-flowing, turbulent, even torrential streams and rivers. Here the sense for pressure waves is bound to break down. If a species of fish wants to conquer this living space for itself, it needs a completely new secret weapon.

The South American knife-fish have succeeded in making a most amazing special adaptation. The members of this family do not, like *Gymnarchus*, fire 300 electric impulses per second. Rather, there are species which transmit only very slowly, perhaps twice a second; others again manage as much as 1,600 impulses in that time, and there are many transition forms between the two extremes. In the rapids regions of rivers there are generally several species living; the slow transmitters make their homes where the whirl is still comparatively gentle, while the fast transmitters go where the waters are at their wildest.

Furthermore, the electric transmitter can also be used as a means of communication. Some species of elephant-fish, which live sociably, use electric signals to keep their school together. But most species of elephant-fish are aggressive solitaries. So they use their electricity for the contrary purpose – to mark off the parts of the river they claim for themselves. A dramatic electric duel between two rivals has been described by Professor Franz Peter Möhres of Tübingen.[22]

> If another male of the species intrudes into an occupied territory, he registers its owner's discharges more and more the nearer he comes to him. But the owner of the territory also discerns the approach of the intruder. As a rule, both sides answer the adversary's signals by a considerable acceleration and strengthening of their own discharges. A struggle of discharges ensues unless one of them withdraws. This quickly turns into a real fight with furious ramming blows and bites.
>
> In most cases, however, the "political high tension" indicated by the threat pattern of the discharges is enough to induce one of the fighters to withdraw. Such

a territorial fight limited to the electric sphere is recorded
in Figure 89. First it shows the discharge pattern of
the owner of the territory. Then the altercation begins.
The discharges soon follow each other so closely that
at this scale the single impulses can no longer be made
out. But the expert notices that gradually the "electric
argument" is turning into a monologue again. In the
end we see the victorious owner of the territory gradually
quieting down.

Here with elephant-fish the discharges apparently
play the same part as the so-called territorial song with
birds. The fish's territory extends just as far as the discharges can be registered underwater, an extremely
practical method of territorial demarcation for creatures
which are active at night and live in waters which are
mostly turbid.

*Figure 89. The electric territorial battle of two elephant-nose fish
is shown here by oscillographs as a series of electric impulses. Further
details are given in the text.*

There are many more things we should like to know.
During the spawning time, of course, the unsociableness
of the elephant-fish must be broken down if the species
is to survive. Is the readiness for procreation indicated
by a change in discharging behaviour? Or are there
perhaps even "electric mating songs"? It would not be
surprising.

But this still does not exhaust all the possibilities. As remarked by Dr. Wilhelm Harder,[23] one of Professor Möhres's colleagues, the electric eel is suspected of attracting fish it feeds on by electric impulses; that is, of carrying out regular electric fishing.

Fishermen can easily attract tuna into the vicinity of fishing-lines and into nets by taking advantage of a pheno-menon in tuna behaviour so far unexplained: the fish swim to a source that is sending out direct current impulses of a particular strength and firing speed. It has not yet been established whether the electric eel also makes use of this modern technological trick and has done so since time immemorial. But it certainly possesses three different types of electric transmitters (see Figure 87). By its low-voltage organ it takes its bearings similarly to the slightly electric fishes just described. The second "living battery" produces electric shocks which in their shape and duration tally to an amazing degree with the values that give the electro-fishermen their best catches.

Finally, a notice might be affixed to the third organ, saying: Beware dangerous high voltage! Depending on the size of the eel, it produces between 300 and 800 volts. As soon as a fish it preys on has come near enough, the eel directs a regular drum-fire of electric impulses at it. Since the prey is a far better conductor than the fresh water all around, it attracts its enemy's electric discharges just like a lightning conductor. The victim immediately starts reeling. Its muscles contract as if in convulsions, quivering and trembling. Stunned and defenceless, it is at the mercy of the aggressive hunter and is devoured forthwith.

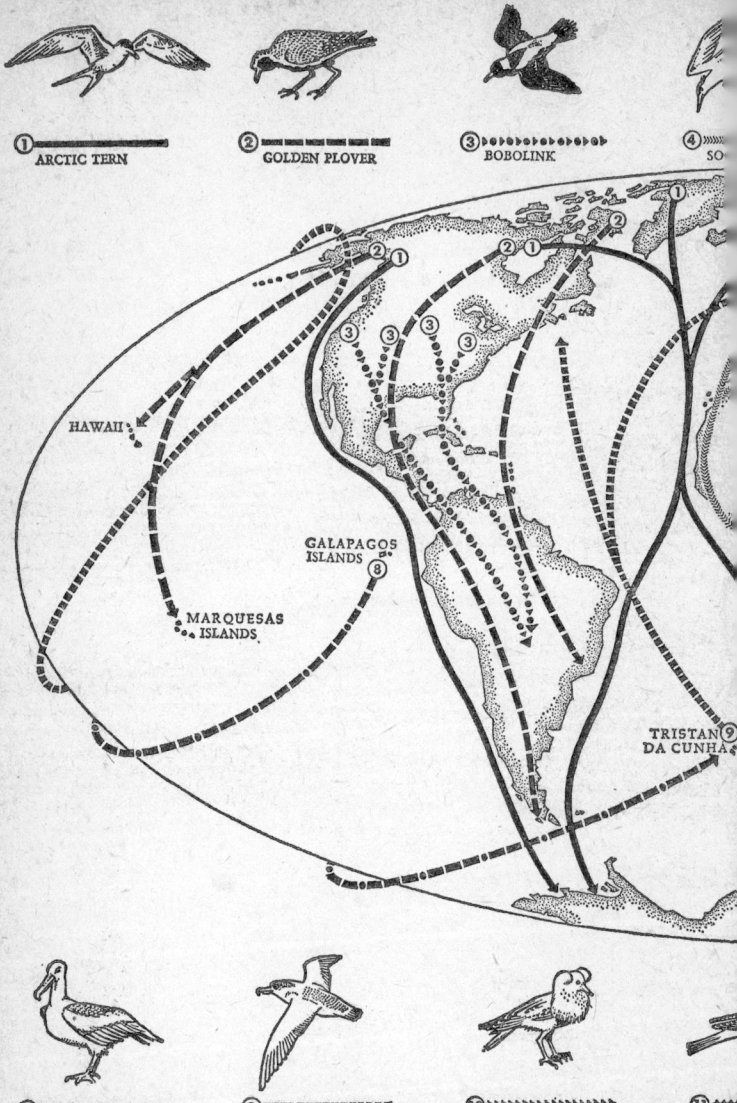

① **ARCTIC TERN**

② **GOLDEN PLOVER**

③ ▶▶▶▶▶▶▶ **BOBOLINK**

④ SO

HAWAII

GALAPAGOS
ISLANDS ⑧

MARQUESAS
ISLANDS

TRISTAN ⑨
DA CUNHA

⑧ **WANDERING ALBATROSS**

⑨ **GREATER SHEARWATER**

⑩ ▶▶▶▶▶▶▶▶▶▶▶▶▶ **RUFF**

⑪ WILI

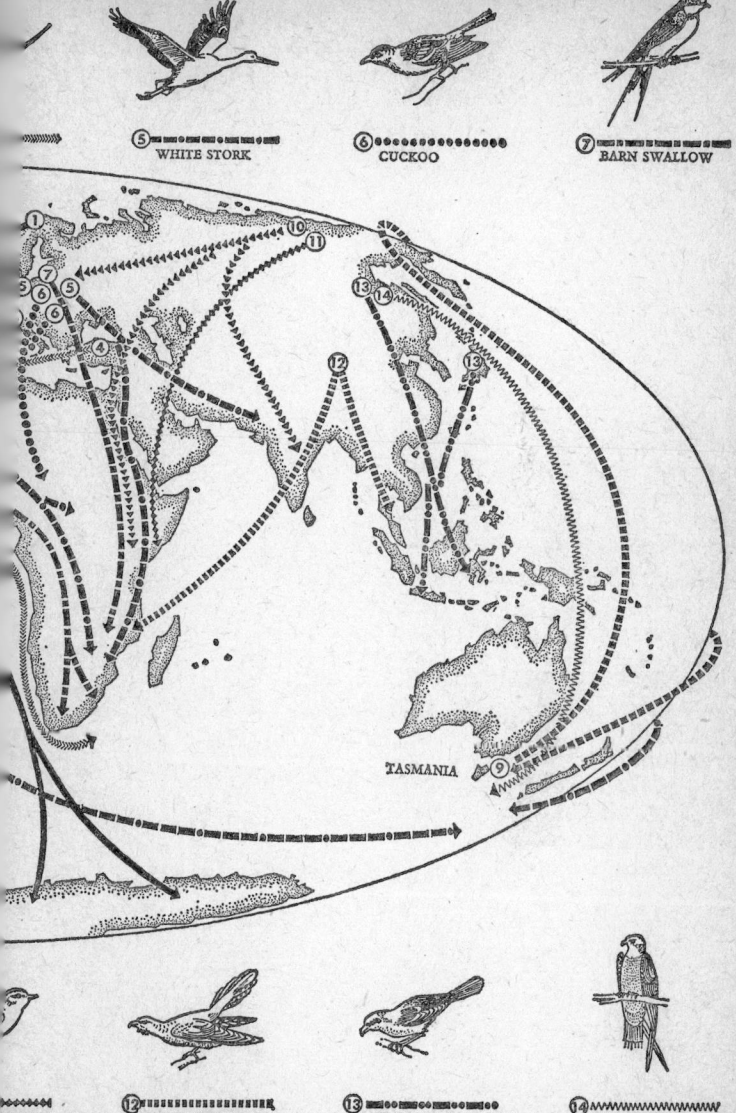

⑤ WHITE STORK　⑥ CUCKOO　⑦ BARN SWALLOW

TASMANIA ⑨

ARBLER　⑫ LESSER CUCKOO　⑬ THICK-BILLED SHRIKE　⑭ SPINE-TAILED SWIFT

Chapter Eight

Migration

Birds

Who would dare travel from Germany to South Africa and back without map or signposts, without ever asking anyone the way, without a travelling companion who knows the route, or anyone showing or telling him beforehand how to reach his destination? This is equivalent to what millions of birds of passage do every year, including finding on return their precise place of birth – say back garden, second oak on left, of house No. X, Y Street, in Z district of a particular town. When we think of the achievement involved, it makes us gasp. But in other parts of the world the phenomenon of bird migration assumes even more amazing forms than in the air passage between Europe and Africa.

While the European cuckoo migrates to Central Africa, its New Zealand relative, the bronze cuckoo (*Chalcites Cuccolus*) leaves a time-table for its young as well. Dr. E. Thomas Gilliard[1] reports that as soon as the female has laid her eggs in the nests of other small birds, she takes up her winter quarters. Her young, lovingly brought up by strange mothers, follow her there a month later on their own and without her guidance.

First their route runs west for 1,250 miles to Australia across an enormous expanse of ocean without islands. After pausing briefly for rest and food in Australia, they continue their journey due north along the whole coastline, and pass over New Guinea to the Bismarck Archipelago. Here at last, after covering a flight distance totalling nearly 4,000 miles, they join their parents whom they have never seen before and of course do not recognize as members of their family. Yet it was these parents who gave them their "flight tickets" even before they were hatched.

Then there is the slender-billed shearwater (*Puffinus tenuirostris*), which every year flies halfway around the Pacific Ocean. In 1798, when for the first time Captain Flinders and his ship's surgeon, Dr. Bass, crossed the straits between Australia and Tasmania now called the Bass Straits, they reported: "The birds passed over us continuously for 90 minutes in a dense flock about 300 yards wide. We estimated their number at 151 millions."

Today's estimates would give about the same numbers, although for some decades many have been shot for canned meat (they taste rather like mutton). At any rate, the flock of shearwaters is enormous. Where does it hail from and where is it bound for? To find this out, two zoologists, John Warham and Dr. D. L. Serventy,[2] ringed 32,000 birds in the early 1960s. They traced the birds' route and established the following round-the-world time-table:

Arrival on the island breeding-grounds between Australia and Tasmania occurs during the nights of September 26 and 27. After some days of courting and mating all the birds disappear again into the expanses of the ocean. But on November 19 the flock, numbering millions, is suddenly back again to lay their eggs in earth furrows. First the male incubates them for a fortnight, while the female fishes the high sea more than a thousand miles away. Then the partners take turns of between eleven and fourteen days on the nest. By the middle of January the young are hatched and are then fed till the middle of April.

Then the great flight starts. The flock of millions first travel 6,000 miles north across the Coral Sea, and by way of Melanesia and Micronesia through the Straits of Korea into the Sea of Japan.

In June the flock shows up in the Bering Sea between Siberia and Alaska, to move southwards along the Pacific

coast of Canada in August. At about the latitude of the
frontier between Canada and the United States, the shearwaters veer to a south-westerly course and cross the entire
central Pacific ocean by way of Hawaii and the Fiji Islands,
to arrive again after a total flight distance of over twenty
thousand miles on the shores of Tasmania punctually to the
very day, on September 26 and 27.

So far it has been assumed that animals and particularly
migrant birds used the day-length as an indication for their
"internal calendar." With the shearwaters, however, which
within a few days pass through latitudes with very different
day-lengths, it does not seem as if this time gauge would
work. At present, therefore, it is still a mystery what calendar
the bird does use so that it can migrate with such precision.

The bird's western relative, the greater shearwater (*Puffinus
gravis*), covers a mere 15,000 miles a year. But it too achieves a
miracle of navigation, starting from Scandinavia, Greenland,
Iceland, and Newfoundland, flying almost exclusively over
the sea, and finding its way to the remote and tiny archipelago
of Tristan da Cunha, in the middle of the South Atlantic,
where it nests with four million of its species. It seems to be
absolutely sure of its route, for it starts its courtship while
still on the high seas.

Eternal summer and perpetual sunshine are the dream of
the Arctic tern (*Sterna paradisaea*), a restless wanderer from
Pole to Pole, from the Arctic to the Antarctic and back again.
Almost as soon as the midnight sun has sunk below the
horizon on the shores of West Greenland and the islands of
North Canada, the birds start out again for the southern
tip of Greenland, then continue their flight to the western
European seaboard, where they join flocks of Icelandic members of their species.

Together they fly on to the African coast. At the latitude
of Dakar the flocks part. Some of them fly on to Cape
Town, then cross to the Antarctic Continent, where by now
summer has started. The rest of the birds fly across the
Atlantic, following the normal air-line route from Europe to
the coast of Brazil. From there they turn south by way of
Tierra del Fuego or the Falkland Islands to the Antarctic
peninsula, where they join other members of their species,

which have started out from Alaska and flown along the whole Pacific coast of North and South America.

It is perhaps not *so* surprising that sea-birds should be capable of crossing oceans; what seems quite fantastic is the carrying out of long overseas flights by birds that cannot swim. I have already mentioned the 1,250-mile flight of the bronze cuckoo, but that is far surpassed by the American golden plover (*Pluvialis dominica*), a bird related to the lapwing (*Vanellus vanellus*), and a little smaller than a pigeon. These birds start out from Alaska on a 2,500-mile non-stop flight to Hawaii. Part of them spend the winter there, but others fly just as far again to the Marquesas Islands in the Pacific.

The Mongolian plover flies annually from Siberia by way of Malaya and the Indonesian Islands to Australia, or over India and then nearly three thousand miles across the Indian Ocean to South Africa. This route, incidentally, is also taken by the lesser cuckoo which breeds in China.

King among the world-tour birds is the albatross, that white oceanic bird with a wing span of between nine and thirteen feet. Albatrosses have been recorded as flying up to fifty miles an hour, without once flapping their wings. This bird is master of a technique which we humans have so far not been able to copy: dynamic gliding. It glides continuously, close to the surface of the sea, using variations in the strength of the wind to provide lift. By using this "dynamic gliding," albatrosses largely avoid the necessity for powered flight.

The albatross flies so far and over such isolated regions of the sea that so far no one has been able to trace its route. But an experiment carried out by scientists for the United States Navy had an astonishing result. On one of the Midway Islands a breeding colony of Laysan albatrosses (*Diomedea immutabilis*) interfered with the flights of a new air base. So the zoologists wanted to transfer the birds to a place from where they were sure not to return to Midway.

As a test, eighteen grown albatrosses were taken by plane a distance of over three thousand miles in various directions to the state of Washington, Alaska, Japan, New Guinea, and Samoa. The experiment was a total failure. From all these Pacific coasts fourteen albatrosses returned within a short time to Midway, the fastest of them after only ten days: a homing feat that would turn breeders of prize-winning homing pigeons green with envy.

The albatross, however, does not hold the speed record. The same speed is attained by the barn swallow (*Hirundo rustica*), which spends the winter in southern Africa; while the European swift (*Cypoelus apus*) is even faster, flying at about a mile a minute. At present the fastest bird in the world is thought to be its East Asian relative, the spine-tailed swift (*Chaetura caudacuta*), which flies from eastern Siberia via Japan, the Pacific Ocean, New Guinea, and Australia to Tasmania. With this bird a top speed of ninety miles an hour has been recorded – although the maximum speed attainable by an individual bird has no relation to its mean speed of migration.

We still know very little about the height at which migratory birds fly. Professor Ernst Schüz[3] has compiled some interesting observations. If the weather is fine, they often fly so high that they cannot even be seen with the naked eye. Several "large birds" (their identity could not be discovered) have been sighted from planes at heights of between 6,000 and 12,000 feet over the North Sea and the English Channel.

After making observations with a small anti-aircraft range finder on the bird island of Mellum, east of the German North Sea spa of Wangerooge, Dr. H. Rittinghaus[4] reported that in that locality he had seen no larks or golden plovers flying higher than 1,200 feet. According to Rittinghaus, lapwings and hooded crows observed there hardly ever rose above 1,500 feet, while jackdaws reached 2,100 feet at most, and rooks 2,250.

Generally this seems to be the most favourable height. From 2,100 feet the birds can theoretically see 75 miles. Also, this is not yet too cold, and the air still contains plenty of oxygen.

But Siberian storks and ruffs have to surmount the snow-capped giants of the Himalayas. It has been reliably observed that in doing this they must climb to a height of 18,000 feet so as to fly through a pass towards warm India.

The Routes of Whales and Fishes

The deep-sea cable connecting the Chilean coastal towns with Peru had been working perfectly for years, and by all estimates would go on doing so for decades. But suddenly the telegraph connection was broken off, and none of the experts knew what the trouble was.

The only thing to do was for the cable-laying ship to lift the precious cable again, starting from the port of Callao. After about sixty miles they found the fault. From a depth of 3,750 feet the ship heaved up a sperm whale, sixty-five feet long, which had the wire wound round and round itself. This proved that sperm whales can dive to depths of the sea in which any normal submarine would be crushed.

What are they doing so far down? In an attempt to find this out, Dr. Hans Hass once devised a bold plan: he would strap a pressure-proof combination of film camera and reflector round the belly of a sperm whale which had surfaced, leave it to dive, and take the exposed film from it when it re-emerged.

A highly dangerous undertaking, remembering the perfectly true tales of the whales of old in the days of the sailing ships. In 1820 the whaleship *Essex* out of Nantucket was cruising in the Pacific among a whole school of sperm whales. A big bull happened to bump into the ship quite unintentionally. She sustained a leak, water started pouring in, and pumps had to be applied. The whale too seemed to have been badly injured by the collision, for it was writhing about in the water as if stunned.

After some time, however, it recovered, and now went for the big ship deliberately in a wild fury, hitting her with such force that her side of thick oak was completely staved in. She sank immediately, and the crew of twenty were left to try to make their way to land in the whaleboats. It was a full three months before eight survivors were found.

Such a fatal collision was by no means an isolated case. An unusually large specimen, caught by a whaler in 1867 without difficulty, had two harpoons still in its body, while the head showed an enormous wound with big fragments of broken ship's timbers sticking out.

Old bulls already wounded were reported to show amazing skill in parrying the attacks of whalers. One succeeded in destroying four harpooning boats in turn by smashing them with its mighty tail fin or crushing them between its jaws. Such experienced solitaries were well-known to the seamen and feared by them as "biting" or "fighting" whales. Some of them, like the fictional "Moby Dick," in Melville's novel, achieved individual fame or notoriety.

One such, according to Erich Dautert,[5] was a gigantic old sperm whale haunting the waters around New Zealand, shunned as "New Zealand Tom." It is said to have shattered many harpooning boats by deliberate attacks and killed their crews. Eventually several sailing vessels pursued it in a joint effort to put an end to its activities. In defending itself the whale had soon crunched up or smashed nine boats and killed four men. The whaling expedition had to give up the fight, and the whale escaped.

This was the kind of monster round whose belly Hans Hass once planned to strap a reflector and film camera! He had an unpleasant surprise on the Azores after asking some native whalers to open a sperm whale's stomach. It disclosed among other things the carcasses of three sharks, almost undigested, the largest of them eight feet long. They had evidently been swallowed by accident. Since after this incident no member of Dr. Hass's expedition felt inclined to face the fate of Jonah in the Bible, he had to abandon his plan.

So for the time being we still have to rely on the remains of sperm whales' stomach contents to help us piece together what is happening in the depths of the oceans. These remains are mostly of squid, which also inhabit the deep sea. Some specimens have a body six feet long, with eyes twelve inches in diameter. The tentacles, twenty-five feet long, are as thick as a man's thigh, and have suckers. An odd favourite dish; the whale that became involved with the cable off Peru evidently mistook it for a squid's tentacle.

All sperm whales have deep scars on their heads where squid have wounded them or torn pieces off their skin with their suckers. About one sperm whale in every sixty caught has a badly distorted spine; the whaler Dr. Hanno Ciliax[6] has deduced that the battles taking place in the dark depths may sometimes even be fatal for the whale, particularly while it is still young.

Squid sometimes congregate into great schools and roam the oceans, also hunting for food. They are pursued by the sperm whales, which is why the whales too embark every year on their oceanic world tours, following a fixed time-table.

Sea captains in the old whaling days from 1800 to 1860 probably knew a good deal more about the sperm whales' routes than our scientists do today. But some of them kept their knowledge to themselves, and took their secrets to their graves; later, the rise of the petroleum industry in America greatly reduced the demand for whale oil.

Between 1955 and 1958 sea captains began taking an interest in these routes again, for the benefit of science. At the request of Professor E. J. Slijper, [7, 8] of Amsterdam, about a hundred Dutch ships' officers (in freighters, passenger vessels, and warships) observed eleven thousand whales during that period. They could not follow up the whales, of course, so the relevant facts from individual reports had to be put together like pieces of a jigsaw in years of laborious synthesis.

The picture eventually built up was this: for part of the year the sperm whale bulls are in a closed male society and follow a different course from the herds of cows and calves. In the Pacific the separate groups apparently unite twice a year in certain spots, which I shall not disclose, as the whales are already threatened with total extinction by whaling expeditions.

A little later the whales disperse again for a few weeks, forming smaller herds, in which bulls, cows, and calves now cross the oceans together. Normally they travel at ten miles an hour. But a whale has muscular strength a thousand times as great as man's, and in case of danger they can drive their great bulks through the waters at nearly twenty miles an hour. Herman Melville, who sailed on several whalers, gives a very impressive description of such a herd crossing the Sunda Strait between Sumatra and Java.

But once a year all the members of a "tribe" seem to meet for a huge, loose "assembly" in the legendary hunting-grounds of the old whalers. We still do not know whether this is due to a simultaneous gathering of great schools of squid in the ocean depths, or to a social urge.

The fin whale, too, seems to know quite well the most favourable routes connecting the Pacific with the Indian Ocean by way of Indonesia. This giant, which may be up to eighty feet long, lives on the teeming crustaceans of the Polar seas, but it also goes in for regular raids on schools of herring, using its wide-open mouth like a net to fish amidst their crowded mass.

Every year it changes its haunts from the Bering Sea to the Indian Ocean. American scientists[9] have observed flotillas of fin whales every spring off the coasts of California and Oregon. They are bound north past Vancouver Island towards the Polar waters which are rich in plankton. In autumn they are suddenly back again to linger a little off the shores of Oregon and California, before eventually disappearing in the expanses of the Pacific.

Whether they are indeed the same whales that pass Indonesia a few weeks later is to be the subject of close investigation. It has been discovered that fin whales accumulate a radio-active zinc isotope in their bodies. So it will be fed to them in larger but still safe quantities in the hope that a research ship with highly sensitive Geiger counters will then be able to keep in constant touch with them and follow them throughout their global route.

Like the big sea mammals, a number of much smaller fish also make fairly extensive sea voyages. One is the North Sea cod,[10] which was formerly believed to stay in the North Sea all its life. But in 1962 this was shown to be an error when fishing experts marked the fins of these fish by numbered metal and plastic discs. True, in February most of them assemble south west of the Dogger Bank, spawn there right into April, and afterwards spread all over the North Sea again. But there are at least some individual specimens which cross the Atlantic to hunt off Newfoundland.

Tuna fish[11] have changed their habits surprisingly in recent times. After the Second World War great schools of tuna suddenly turned up in the North Sea and the Kattegat, and many herring-catchers changed over to fishing for this valuable catch. Professor Meyer-Waarden of Hamburg discovered the course of this unexpected windfall.

These giant horse-mackerel, which reach a length of six feet, set out from Morocco and Spain in April and May, and go first to Sicily to spawn there. Then they leave the Mediterranean by the Strait of Gibraltar, pass the west of Ireland, round the northern tip of Scotland, revert to a southern course, and arrive in time for the summer herring season at the Dogger Bank. Later they return by the same route to the Mediterranean.

Some years, however, a transatlantic "population exchange" takes place.[12] For reasons still obscure tuna fish suddenly transfer in masses from the European to the American shores of the Atlantic, or vice versa. There seem to be "cosmopolitan" tuna in the Pacific as well, for fish from California have been traced on the shores of Japan.

Since 1956, just as inexplicably, fewer and fewer tuna have come to the North Sea. Today tuna-fishing has practically ceased in these waters, and the fishermen are trying to return to their herring interest.

The herring are also setting us problems, though. With them too, catches have gone down so much since 1959 that Norwegian fishermen are worried for their livelihood. The reason, however, does not seem to be that the stocks have been overexploited, but that there is a strange cycle, which takes place roughly over a century, in the route of the herring migration.

Dr. G. Hempel[13] of the Kiel Institute of Marine Biology writes:

> By the use of research ships, submarines and aircraft, and the marking of several hundred thousand herring, the migration of Norwegian herring between the feeding-grounds near the Polar front north and east of Iceland and the spawning-places on the Norwegian shores over six hundred miles distant has now been fairly well

explored. After the feeding period in summer the herring
withdraw for several months into the cold water east

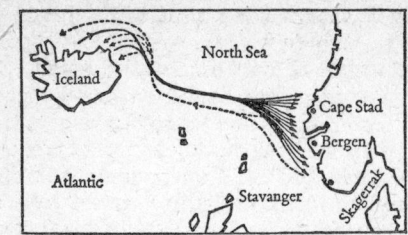

*Figure 91. The migration routes of Norwegian herring lead from the
feeding places near Iceland to the spawning places on the coast of
Norway. But the spawning places change by laws so far unknown. At
the end of the last century they were in the Skagerrak. At present they
are to be found north of Cape Stad. This cycle is repeated about every
hundred years.*

of Iceland for a winter rest. In January they set out to
cross the north-east Atlantic, proceeding eastward at
depths of more than six hundred feet. Whereas normally
the herring schools disperse at the fall of darkness, the
herring here keep together even without light, covering
up to forty nautical miles a day.

Their destination is also subject to strange changes.
According to Dr. Hempel:

the catching period now finishing started at the end of
the last century with ample catches in the Skagerrak. In
the course of time the spawning-place at the south
coast of Norway moved more and more to the west and
north. After the First World War the main spawning
activity shifted from the south to the west coast into
the waters round Stavanger. In the Second World War
it took place round Bergen, and for several years now
it has been the coast north of Cape Stad near Aalesund
and Kristiansund. Simultaneously with the shift in
locality, spawning has started later, from between
December and January to early March.

Old records show that in past centuries quite similar developments occurred. In fact, F. Devold, the Norwegian marine biologist, has predicted optimistically that the present recession, far from heralding the end of herring catches, means that a new boom will soon begin.

Formerly, too, the catches became a good deal poorer as soon as the spawning-places shifted from the Kattegat to Cape Stad: "For a few years fishing for fat 'large herring' near the Lofoten Islands provided some compensation. But the hard times for the Norwegian herring fishers after that were brought to an end only by a revival of herring catches in the inner Skagerrak and on the south coast, which amounted to the start of a new catching period."

Why the Norwegian herring changes its spawning regions in a cycle of roughly a hundred years is still an enigma to us. As history shows, human fishing is apparently not to blame. Perhaps the reason is connected with other gluttons for this tasty fish, like the fin whales, cod, and tuna, already mentioned, and the spiny dog-fish.

This shark, which abounds in the North Sea, is only about three feet long. When sold by fishmongers, its two dorsal fins with their offending spikes have been removed. Nor do people who eat it have to worry about indirect cannibalism, for it never attacks bathers or survivors of shipwrecks, but only herrings. .

Like human fishermen, the spiny dog-fish also likes to be present wherever herrings congregate in great numbers. For this purpose it makes long journeys, especially along the Atlantic coast of North America. As Professor Walter N. Hess[14] of Converse College, South Carolina, has established, every spring the dog-fish, in schools often up to a thousand strong, swim a distance of 1,500 miles from the coasts of the Carolinas and Virginia, to Labrador, and in autumn return to their winter quarters.

If we still have no proper explanation of how whales can find the Sunda Strait, tuna the Strait of Gibraltar, or herring their spawning-places, the travelling route of the eel at present seems to us even more mysterious.

Ideas about this used to be relatively simple. Eels spawn in the Sargasso Sea, that part of the North Atlantic between

the Azores, the Bermudas, and the West Indies which is full
of floating seaweed, particularly *Sargassum bacciferum*, from
which the sea derives its name. Currents here round up the
seaweed, as it were, so that it cannot be washed away. Sometimes it covers large reaches of the sea so densely that ships
avoid these reaches for fear of getting stuck in it.

Here the tiny eels are born. Somehow they break free of the
Sargasso Sea and then let themselves drift with the currents.
Some of them are carried to America, others to Europe,
floating with the Gulf Stream. For this 3,000-mile journey
they need three years. As elvers (young eels) three inches
long, they start their ascent of the rivers in spring. By way of
the Rhine, they even reach Swiss mountain ranges up to a
height of 10,000 feet.

In these inland waters the males stay for between three
and five years, the females for nine years. Then they leave
the European rivers as large, fat, glossy silver eels. What
happens now has since 1959 been the subject of vigorous
controversy among scientists, and there is so far no definite
answer. For nobody has yet been able to follow the adult
eels' route in the open sea.

First of all it was taken for granted that they returned to
the place where they were born, the Sargasso Sea, to spawn
and then die. But the trouble with this assumption is that
when they enter the sea, their anuses close up. From now on,
without any intake of food, they have to swim 3,000 miles
against the Gulf Stream for about a year, living only on
their internal fat resources.

Their store of energy is not nearly enough for such a vast
effort. This was worked out in 1959 by Dr. D. W. Tucker,[15]
who accordingly takes the view that the European eel perishes
somewhere at sea. He therefore calls it "only a useless waste
product of the American eel," which can find the Sargasso
Sea much more easily and reach it faster.

But oceanographers[16] have since discovered that below
the Gulf Stream there is a counter-current flowing at a
greater depth east-west, from Europe to the Sargasso Sea.
Perhaps this speeds the eels on their hard journey like a tail
wind, and also acts as a signpost.

Even if that were the case, it would still be a miracle for

the eels to find their far-off destination. Imagine a diver plunging down into the depths of a river, seeing nothing round him for months but water, a bright glimmer above him, a dark abyss below, feeling completely alone at night, fleeing from enemies by day. How could he ever find the Sargasso Sea?

Insects

One day in January the middle of the Masai plains in what is now Tanzania was a busier thoroughfare than Piccadilly Circus or Times Square, not for human traffic but for swarms of flying insects in their millions.

Professor Carrington B. Williams,[17] who was an eye-witness, gave the following description:

> The very definite way in which some species of butterflies and locusts keep to their self-determined course is also illustrated by observations of different species migrating simultaneously in the same locality but in different directions. I saw a particularly interesting case of this in north-east Tanganyika during January 1929 when, in the sunny hours of every fine day between 11 A.M. and 3 P.M., there were two constant streams of butterflies; one the large yellow *C. [Catopsilia] florella* going to the north-north-east, and the other a small yellow *Terias senegalensis Boisd.* going to the south-west.

These butterflies crossed through each other's flight paths in an area about 30 yards across, Professor Williams writes, during

> periods of observation on various days. The problem was made even more complex by the fact that on 29th January there was a flight of millions of locusts – *Schistocerca gregaria* – going to the south-east. The different insects, nearly all below ten feet from the ground, interlaced in their flight, and the movements of one species had no apparent effect on the direction of the others.

These insects must be envied their ability to react like lightning so as to avoid bumping into other travellers. Migratory birds which move in dense flocks have the same capacity. In a matter of seconds a swerving movement spreads from the front to the back across tens of thousands of birds. At present it is not clear whether this phenomenon is due to reactions spreading almost directly from one bird to another, or whether special command calls are also involved.

In any case no collisions, sudden tumbles, or even reeling unsteadiness, have ever been observed in a flock of birds, despite the high speed with which they shoot along and carry out their manoeuvres. How superior in this are the swallows and starlings, compared to human motorists. Biologically, the trouble about man as a driver is that his reaction time is adapted to a pedestrian speed. Pedestrians among themselves have as little need of traffic rules as butterflies, locusts, and birds, and they collide as rarely as these, unless somebody is not looking where he is going.

As everybody knows, traffic rules cannot adequately compensate for the driver's shortcomings, biological and otherwise. So laws and penalties, even if they are gradually made more and more severe, will not achieve a decisive reduction in accidents. The only real remedy would seem to be the development of artificial sense organs reacting with the speed of swallows, like radar-controlled brakes and other motorists' accessories (see "the electronic frog's eye", p. 29).

To return, however, to the teeming traffic of insects above the Masai plain: where do these swarms of millions of butterflies come from and where are they bound for?

It is only during recent years that zoologists have started investigating certain phenomena in the insect world which have surprising analogies to bird migration. Butterflies fly distances of 2,000 miles; moths cross oceans in vast swarms; hover-flies and mosquitoes negotiate mountain chains by way of passes; ladybirds and dragonflies cross whole continents.

In some years the flight of the painted lady leads across the Mediterranean. On a ship thirty miles off the North African coast Professor Williams watched millions of these butter-

flies flying in a broadly dispersed front of a hundred miles
on a northerly course. They were probably the same swarm
which was later seen crossing the Alps. The valleys funnelled
the loose formations into compact columns. When they had
reached the plains north of the mountains, a social integrat-
ing force still kept them close together. Thus in the open
country near Lake Neuchâtel they were heading north still
in a procession between ten and sixteen feet wide, which
took two hours to pass.

In years of unchecked mass reproduction, cabbage whites
from the Baltic move across the North Sea to England in a
thick cloud like drifting snow. The hawk-moth, living in
tropical Africa, with the odd capacity of producing squeaking
noises when in trouble, also flies over the Mediterranean, to
reach France and Germany, Sweden and Finland, sometimes
even Iceland. The scientists of a British weather ship[18] in a

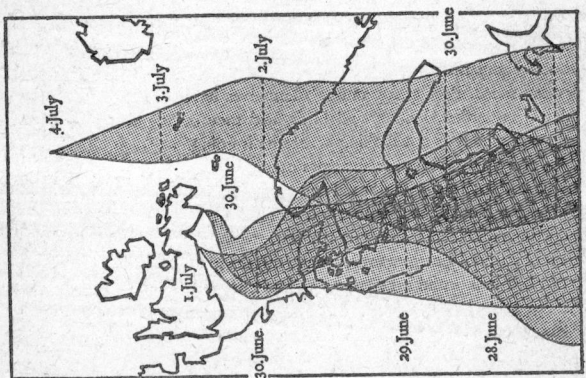

*Figure 92. Mass swarms of Diamond-backed Moths were sighted at
the times mentioned in 1958 at two places on the North Sea coast of
England and by a weather ship on the high seas. The dotted areas
show the air masses that carried the moths with them. They probably
came from the region in which the three dotted areas overlap.*

sea area west of a line from Iceland to Ireland watched a
swarm of tiny diamond-back night moths, which two days
before had left Norway, nine hundred miles away, in a

strong east wind. Whether the swarm ever reached America is not known.

The most magnificent feat is achieved by the copper-coloured North American monarch butterfly. It spends the winter in California, Mexico, Louisiana, and Florida. There are sometimes thousands of them sitting half asleep, in a hibernation-like state, always on the same "butterfly trees": one of the sights for tourists to marvel at.

With the beginning of spring, about the end of March, the butterflies wake from their numbed state. First they fly about a bit in small swarms looking for food. But gradually individual insects slip the swarms to go off in a northerly direction.

In years of moderate population density, then, it is only individuals that cross the whole continent, right from the Atlantic to the Pacific. Two months later they reach Canada and Hudson Bay. Man is not conscious of the individual migrants, for you see only a butterfly now and then, and unless you are a collector take no special notice of it. But entomologists who have caught marked specimens (with stamped wings) could reconstruct their exact route.

There are also periods, however, of apparently boundless mass reproduction. Then the monarch butterflies make the long journey in swarms. A report from the year 1885 relates that one swarm settled for a rest in northern New Jersey, in an area six hundred feet wide and two miles long, closely covering the branches of every tree, wing to wing.

From September onwards the butterflies return again to the warm South. Although they are not the same individuals which set out north in spring, they roughly know the home of their previous generation. But among the individuals that fly from Hudson Bay to Florida in the autumn, it has been established that quite a number, after spending the winter there, go back again to the far North the same way next spring. So they cover during their lives a distance of at least 2,500 miles.

Scientists have observed a distinct tendency to fly north in spring and south in autumn, in other species as well, such as the red admiral, sulphur, large cabbage white, and small postillion butterflies, the white pirate moth, and some night moths. Our knowledge of this insect migration is still very incomplete, but it is already becoming clear that these migratory flights are far from being exceptional or accidental phenomena.

The exacting and dangerous journeys are certainly not prompted by shortage of food or a change for the worse in living conditions. Monarch butterflies, ladybirds, hover-flies, and night moths start returning to their winter quarters a long time before the cold weather sets in. As with migratory birds, there is a strange urge to be on the move, which is inherent in the species, forcing the insects to leave the place where they are and fly off, keeping to a certain flight direction. But whereas birds' migratory restlessness is always aroused through hormones by an "inner calendar" during the same annual period, many butterfly species turn into migrants only in years of great over-population. As with locusts and lemmings, it seems to be a scent substance or a kind of psychological stress which triggers off the travelling urge.

First indications of this were observed by Dr. K. Burmann[19] with mass flights of larch leaf-rollers (*Coleophora laricella*) in the Tyrol.

> At a control catch at the end of June 1964 in the region of Trentini, hundreds of thousands of these little night moths made for the 500-watt light trap, so that their approach looked like a dense snowdrift. Moths were moving about in a layer an inch thick. They were pre-dominantly females, incidentally, and still possessed their entire stock of eggs. These females instinctively avoid depositing their eggs on larch branches ravaged by caterpillars, and they form into migratory swarms to set out for districts where the larches are intact.

But this explanation suffices only for the triggering off of the migratory urge: at most, to find suitable habitats very close to the unfit regions, not for the impulse to cross oceans and continents, as with the monarch butterflies. Probably we have to accept that there is some compulsion which makes insects, like birds and fishes, set off on journeys that seem so much longer than is necessary. The urge is as much part of

their lives as their possession and use of heads or l... and
it must have served some purpose, at least original Dr.
C. G. Johnson[18] has perhaps found a clue to that pur... in
the dragonfly.

Among the 4,500 species of this order some are ma...y
local. They live along rivers and even defend their...y
hunting-grounds against rival members of their species...
vagabond dragonflies behave quite differently. In the autu...
for instance, they fly to Spain by the thousands across...
mountain pass of Gavarnie in the Pyrenees (5,000 fe...
They live by pools; that is, near water which may periodica...
dry up.

Permanently localized dragonflies could not settle by su...
pools at all, for the first drought would kill them off. Lookin...
for other pools very near would hardly help them, for in ...
longer drought these would dry up as well. So the only
thing for them to do is to get as far away from the spot as
possible, a hundred miles or even more, without stopping
on the flight to look around. Then they at least have some
chance of reaching an area where rain keeps the pools filled.

So dragonfly species with an innate migratory urge had to
develop before dragonflies could occupy a new living space,
a new ecological niche. The butterflies' migratory urge may
provide similar advantages for the survival of the species.

Incidentally, for many insect pests the agricultural practice
of rotating crops is equivalent to the drying up of pools. So
it is not surprising that they too are distinctly migratory
insects. Sometimes they arrive literally overnight from far
distant regions, without our at present having any exact
knowledge about the laws governing their migrations.

But how does the butterfly, wanting to fly north in the
spring, know where north is? How does the golden plover
of Alaska find the distant island of Hawaii; how do whales
and fish navigate in the open sea? I shall discuss this question
in the following chapter.

Chapter Nine

Animal Navigation

The Sun Compass

One might think that a free animal would make use of its freedom to run, fly, or swim sometimes to one place, sometimes to another. But this is not the case. Many creatures of land, water, and air are governed by the irresistible urge to set out in a definite direction and to keep moving in that direction only. Moths cannot stop themselves from fluttering into the light, where they will burn themselves. Bees are guided straight to the source of nectar and back home again. The same urge controls water-fleas, shrimps, and dung-beetles.

The water-fleas[1] swim directly towards a source of bright light; that is, towards the sun in the day or a bright lamp at night. If they are not diverted – perhaps by the wall of an aquarium, an obstacle, the lure of food, or enemies – they cannot help doing this, and always orient themselves so that light falling on the compound eye is coming from straight ahead. If undisturbed, they will thus describe a semicircle in a lake in the course of a day. In the morning when the sun rises, they swim east, then steadily follow the sun's course, swimming south at noon and west in the evening.

Such behaviour seems rather pointless to us. Yet the water-flea's ability to sight the sun firmly with its compound eyes and keep to a definite course displays in primitive form one of the most amazing phenomena in animal orientation: the sun compass. The albatross, homing pigeon, and salmon are distinguished from the water-flea in this respect only by improvements of the sighting principle.

The first advance is shown by a minute crustacean an inch long of the order *Mysidium*,[2] which can maintain two different sighting angles to the sun: straight towards the sun like the water-flea, or at right angles to the direction of the sun.

According to Dr. Marianne Geisler,[3] the next stage of improvement is represented by the dung-beetle, an insect of interest to the zoologist in many ways. It can sight the sun at three different angles: at 0° and 90° like *Mysidium* and at the intermediate stage of 45°. It has no free choice, though, between these three directions; depending on the time of day, it is compelled by its internal clock to favour either one angle or another.

In the early morning it instinctively "prefers" to see the sun straight ahead, so it steers east. About nine o'clock its mood shifts abruptly: it aims to sight the sun at an angle of 45° in front, using its right eye. Towards noon the sighting angle again increases abruptly to 90°, so the beetle walks in such a way that the sun is directly to its right.

But then all of a sudden the left eye becomes the source of these predilections instead of the right eye. The insect suddenly swerves to a reverse course. At 3:00 P.M. there is yet another change of mood. Now the sun must no longer be straight to the left, but has to shine obliquely from left front at an angle of 45°. In the evening there is a last abrupt change of course. Just as in the early morning, the beetle makes direct for the sun; that is, it heads west.

If one charted all these courses like a ship's officer at sea, one would find that in the morning the beetle heads almost straight east and in the afternoon almost straight back west. But only almost! Closer study reveals curves and corners as well, curves due to the sun's movement and corners caused by the sharp changes of course.

Again we have no idea of the purpose behind this compulsory course, which makes the insect in its movements a slave of the sun and the time of day. In any case we can here see the outlines of a considerable improvement in animal powers of orientation: allowing for the sun's movements according to the time of day, while steering by the sun's position. This is the decisive transition stage between orientation by angles and by compass.

We can now imagine further refinements taking place during evolutionary history. Other insects made smaller and smaller sudden shifts of course more and more often in a day. Eventually there were insects which would correct their course roughly every ten minutes by only 2·5°; that is, by an angle corresponding to the one between two adjoining ommatidia of a compound eye (see Figure 24, p. 52). This practically guarantees a dead straight course, not distorted by the progress of the sun.

We could visualize the phenomenon roughly like this: the insect has an internal clock,[4] with an hour hand that moves round the dial every twenty-four hours; that is, only half as fast as on our mechanical clocks. Now the insect, say a bee flying home, chooses a straight course. The direct sunlight then falls only on a single one of the 5,000 ommatidia in its eye. At first the bee need only fly on so that the sun stays in that ommatidium. After ten minutes the sun will have shifted by 2·5°, but so does "the hand" of the bee's internal clock. So it now sights the sun with the adjoining lens. This change is repeated every ten minutes, and so the straight course is maintained.

The sun compass is of vital importance to bees, wasps, hornets, bumblebees, ants, dragonflies, and other insects which have to find their way home after excursions to unknown regions. It is also used by water striders, wolf-spiders, sand fleas, and beetles, in some of these cases without our having an explanation for it.

If water striders (*Velia currens*),[5] for instance, are put on dry soil and startled, they will always take flight in a southerly direction. No matter where they come from, whether they live on a large lake or a small pool, they all know only one escape direction, and that is south – even though the saving water may not be that way at all.

There seems more point in the compass navigation of the sand fleas, studied by Professor L. Pardi[6] at the University of

Florence. This little crustacean, well-known to bathers on Italian beaches, is in an awkward position if carried onto dry sand by a storm or swept into the sea by waves. From its position on the ground it is generally unable to see its habitat, the shore right by the sea. So it is then reduced to taking its bearings from the sun.

The sand fleas of the Italian west coast near Pisa, for instance, always escape in a westerly direction if conditions are too dry for them, in an easterly one if it is too wet for them. In their normal geographical area this is very practical, for they always find their way back to the moderately damp shore. But if taken by a zoologist across Italy to the Adriatic coast, the course which will ordinarily save them leads them to their doom, because the coast is now west, the sea east. When swept out to the water, they will flee farther and farther from the shore, while the ones left on dry sand hop farther and farther inland till they perish.

If sand fleas from the Adriatic are taken to the Ligurian coast west of Pisa, they will show the equivalent, fatally mistaken reactions. So it is safe to assume that the course of flight is innate in local populations of the species.

Rigid instinctive reactions, conducive to survival in many situations but disastrous in changed environmental conditions, may be regarded as a sort of intelligence substitute provided by nature anywhere that sufficient brain power is lacking. One creature may carry out an action purely by instinct and with only a few innate nerve circuits, while another has to muster all its powers of learning and adaptation, needing a thousand times more "grey matter."

It is interesting, therefore, to find among lower orders a creature able to adapt its sun compass and escape course to changed conditions: the wolf-spider. This arthropod, which does not make nests but pounces on its prey with an enormous leap like a wolf, lives on the shores of lakes, and like the water striders can walk about on the water's surface to catch insects there. If in danger it darts back by the shortest route to the shore, where it hides under stones or grasses.

The Italian zoologists Dr. F. Papi and Dr. P. Tongiorgi[7] transferred some wolf-spiders from one of the lake's banks to the opposite one. Unlike the sand fleas, the spiders did not take long to orient themselves and remember to make for the opposite direction from before if they meant to flee.

The two zoologists managed to deceive the spiders in another respect. They screened the sun by big pieces of cardboard, then reflected the sun "round the corner" so that it was no longer south but north in the spiders' sight. When fleeing, the spiders promptly ran far out into the lake till the sun's true position became visible again from behind cardboard and mirrors. At this point they immediately turned and rushed back to the bank, where they again came under the effect of the trick, so they constantly scuttled backwards and forwards until they were exhausted.

Another strange thing the experiment revealed was that they were still utterly confused, although they must have seen the tall trees on the bank. When the sky is overcast and they have no opportunity of taking their bearings by the sun's position, they unhesitatingly make for the outline of their home bank. Apparently the use of the sun compass makes it impossible for the insect to resort to other landmarks. Here for the time being the power of the instincts leaves no scope for turning meaningless into meaningful behaviour.

The sun compass reaches its highest perfection among insects with bees and ants. On long reconnoitring journeys, while constantly zigzagging and curving this way and that, these insects integrate all the angles and distances of their course into a resultant, so that at any given time they know the direction of home and can make for it directly. This astonishing capacity has already been described in detail on page 54.

One fact, however, has so far been disregarded. Insects sight the sun with their compound eyes in such a way that its daily round must necessarily present considerable fluctuations to them. Let us have a closer look at the point.

Professor von Frisch[8] discovered this when checking whether bees were diverted from their flight and dancing direction when shown an artificial sun in the correct direction but at a different altitude from the real sun. It turned out that for the bee it does not make the slightest difference whether the sun's height above the horizon corresponds to the actual time of day or year or whether it is 25° higher or 40° lower.

"For the bee's correct orientation by the sun, therefore, the image of the sun does not need to fall into the same ommatidia," wrote Professor Frisch. "But it must fall into ommatidia which are in the correct azimuth. Ommatidia looking out on the same direction seem linked to each other in their nervous circuits."

So the bee does something which may appear unnatural to many pupils studying three-dimensional geometry: it projects the sun's course onto the horizontal plane. The height of the sun is completely unimportant to it. All it is interested in is the angle of the sun in the horizontal line, the azimuth.

During the European summer, when the sun rises steeply in the morning, it changes its azimuth only slightly. But at noon it moves almost parallel to the horizon and in so doing

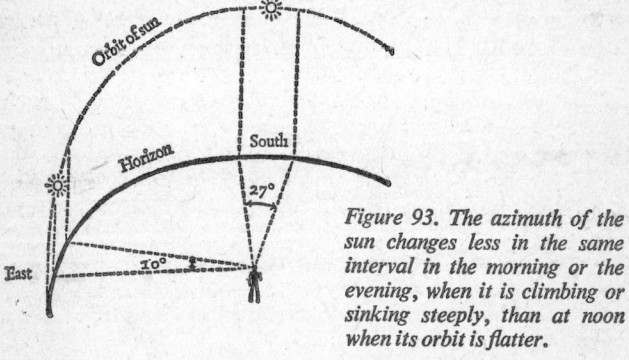

Figure 93. The azimuth of the sun changes less in the same interval in the morning or the evening, when it is climbing or sinking steeply, than at noon when its orbit is flatter.

reaches its highest rate of change of azimuth, which towards evening gradually decreases again.

The sun's azimuth, therefore, does not proceed with the regularity of a clock hand from the east through the south to the west. Moreover, the daily differences in angular velocity change with the seasons and geographical latitude the bee happens to be in. On the sun's southern hemisphere the sun even circulates in the opposite direction, anti-clockwise from the east through the north to the west.

In view of these complications one would not expect a bee's brain to be capable, in orientation, of taking into account the continual variations; that is, by attending to the time of day in relation to the sun's positions. Conceivably, of course, the bee neglects these refinements and behaves as if the sun's azimuth changed as regularly as a clock hand. But during a journey of several hours, such refinements accumulate to errors that would make it impossible for the bee to return home.

In such vital matters nature performs more miracles. Experiments by Dr. J. Reimann,[9] a zoologist of Freiburg, show that both ants and bees can incorporate the changes in the sun's horizontal angular velocity with astonishing accuracy into their calculations.

One morning in June 1964 a group of ants with their broods, fifty-eight insects in all, were making for their hill, steering by the compass, keeping at exactly 90° right of the sun. At eleven o'clock sharp Dr. Reimann enclosed them all in a container, and turned them loose three hours later in a laboratory which had a lamp as a substitute for the sun.

The ants no longer kept to their original angle. Their internal clock and "knowledge" of the azimuth change told them the sun meanwhile must have proceeded by another 84°. From now on they kept only to 6° right of the sun. The sun had in fact progressed by 86° during those three hours.

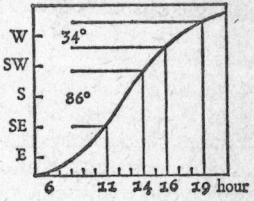

Figure 94. From the azimuth curve the direction of the sun at each time of day can be seen. The example given in this sketch applies to the end of June at Freiburg in Breisgau, Germany. In the three midday hours from 11:00 A.M. to 2:00 P.M. the azimuth changes by 86°, in the three evening hours from 4:00 to 7:00 P.M. only by 34°. The remarkable thing is that certain animals can allow for these differences in their sun compass.

So the ants' error of calculation was the harmlessly small one of 2°.

Two days later another group of ants from the same hill were also locked up for exactly three hours, from 4:00 till 7:00 P.M. As can be seen from the azimuth curve in Figure 94, during this period the sun's position changes only by 34° – a great difference from the 86° in the middle of the day. The forty-seven imprisoned ants, on release, reduced the angle of their course to the right of the sun by 39°. By this adjustment, which was completely adequate, they confirmed the findings of the previous experiment and also showed that they can appreciate how much faster the sun's azimuth alters at noon than in the evening.

Now Dr. Reimann started constructing a regular "sun planetarium" for his ants. There were different circular courses along which a 500-watt bulb could be moved, simulating now autumn or high northern latitudes, now mid-summer or a tropical environment. The ants, it turned out, grasped very quickly any possible course of the sun, and took it into account when determining their own course. Even when Dr. Reimann made the artificial sun take an opposite course so that it most unnaturally rose in the west and set in the east, the ants learnt very quickly to use it as a means of orientation.

To appreciate the ants' feats fully, we need only think what confusion would be caused among human navigators on the high seas if the sun were suddenly to rise in the west and set in the east, so that all tables and manuals were useless. But the ant adapts itself to the change within a few hours. The orientation powers here, then, seem to be based on genuine learning processes, not exclusively on automatic instinctive reactions.

In 1964 a manufacturer of electronic computers exhibited a magnetic memory in a ferrite ring so small that it could be hung round an ant's antenna. Certainly a masterpiece of technology – yet what reckoning feats the ant achieves with a brain much tinier still!

The Perception of Polarized Light

A clear, cloudless sky does not look uniformly blue for the bee with its compound eye, but covered all over with patterns of alternating light and dark. These are by no means optical illusions, they are realities which we cannot see: the pattern of polarized light in the sky registered by more than two thousand separate eyes.

What is polarized light? Light rays are vibrations like the waves of the sea. Only they do not vibrate, like water, exclusively in one plane (up and down), but in all directions at right angles to the ray. This alters when light is reflected by shiny surfaces or is diffused by the earth's atmosphere. Then it vibrates more or less in one direction like the waves of the sea. It is polarized.

Sunlight, which comes to us direct, is not polarized, therefore; but the diffuse light of blue sky is. As can be seen from Figure 95, light from each point in the sky vibrates in

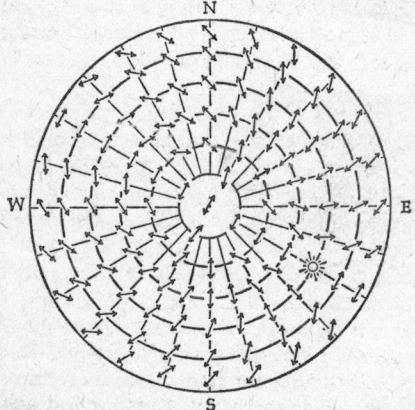

Figure 95. The polarization of sky light about 10:00 A.M. The double arrows show the plane of polarization. The pattern alters characteristically with the course of the sun. Insects know these connections and can work out from them the sun's position, even when it is hidden behind clouds.

a different direction, which changes continually through the day according to the sun's position.

Anyone looking at Figure 95 might suspect that these patterns of light and dark cover the heaven as alternating light and dark stripes, as a sort of contour marking like a mountain range on a map, with the summit as the – unpolarized! – sun. This hypothesis is very tempting and plausible, but it is pure speculation, of course, whether the bee really synthesizes the light-and-dark to such "sun contours." So, to be on the safe side, let us go on talking of light-and-dark patterns. What they look like to the bee, we do not yet know.

Perception of this pattern of polarization in the sky is just as vitally important to the bee as is the sun compass. For in countries like Britain, where the sun is all too often hidden behind clouds, it would be fatal for the foragers if they immediately lost their bearings completely whenever a small cloud hid the sun. So they use two other aids.

One is the sun's ultra-violet rays, which can still penetrate through very thick clouds. Long after man ceases to be able to tell where the sun is, the bee still sees our daytime sky as we register it through a slight mist or a thin veil of high cloud: as a haze with a brighter patch. This has already been described in detail on page 73.

But even the ultra-violet rays do not penetrate through a dense cloud mass; so a further auxiliary sense is needed. As early as 1948 Professor von Frisch[10] had established that any patch of blue sky with a diameter of only twenty to thirty times that of a full moon is quite enough for bees to find their way in.

This was a phenomenon which at first sounded rather mysterious, because we could not grasp the signpost element in a small bit of blue sky. Other researchers thought insects had the ability to recognize stars by day and take bearings by them. But right from the outset, deep insight into physics made von Frisch suspect that the signpost and sun-substitute could only be the polarized light of the sky, the plane of which bees could apparently somehow register.

After the Second World War, when American industry produced polarizing filters as a protection for motorists against glare, the time had come for a series of exciting experiments. Professor von Frisch tipped an experimental hive on its side so that the bees had to perform their dance on a horizontal comb. This would give them a sight of blue sky through a stovepipe set up like a telescope. Successful foragers returning to the hive would then do their tail-wagging dance with the straight part of it pointing exactly in the direction of the field of blossoms they had just visited.

Then von Frisch put a polarizing filter over the top opening of the stovepipe, arranging it so that it only let through light in the plane of polarization which the bees in the hive had been able to see in the sky through the pipe even without the filter. So there was practically no change for the bees, and they continued to point the signalling part of their dance directly towards the source of nectar.

Then von Frisch rotated the filter in front of the pipe, thereby also altering the plane of polarization of the light falling into the hive. Thereupon the dancing bees, when doing the signalling part, turned like puppets in the same direction and at more or less the same angle. So they pointed in a wrong direction, which should have been the right one according to the polarized light.

Through filter and stovepipe the bees in the hive can be shown any direction of polarization, and they will always adjust the signpost part of their tail-wagging dance accordingly. From the light polarization they deduce where the invisible sun is at the time, and they then steer by this "theoretically" determined reference point. No one could really believe it if it had not been proved beyond doubt by many experiments!

If bees are shown a plane of polarization which does not exist in the natural sky at that time of day, the dancers are completely bewildered and disorientated. Some hours later, however, when that plane appears in the sky, they at once work out from it the appropriate sun position, and their dance becomes directional again. So they must somehow have an idea of what pattern the sun shows at any given time of day.

When confronted with a plane which occurs simultaneously in two different parts of the sky, the bees in the experiment develop split personalities. Since two different sun positions correspond to this one plane, their dances in continual alternation point in two different directions.

This case of ambiguity can also occur in a bee's normal life with a single patch of sky which is just visible. An experienced forager, however, is not confused by it, for well-known landmarks like trees, houses, and roads enable it to come to the right decision immediately between two possible courses.

To recapitulate, all this means that from early morning to late in the evening the bee knows the pattern of polarization belonging to each sun position. The linking in the bee's mind of two mutually dependent phenomena makes the sun in any case still the reference point of all the bee's feats of orientation, no matter whether it can be sighted directly or whether its position has to be inferred from other clues.

This perception of polarized light, an achievement of the senses beyond our grasp, has since been established for many other insects as well: wasps, bumble-bees, ants, flies, caterpillars, beetles, water striders, water-fleas, funnel-spiders. It has been established also for squid and octopus. This is particularly remarkable in the latter, for these molluscs have no compound eyes or ocelli, but camera eyes resembling our own.

What analyser gives them all this extraordinary ability? The first suspicions were directed to the lenses. Each of the 2,500 separate eyes in the bee's compound eye has a lens.

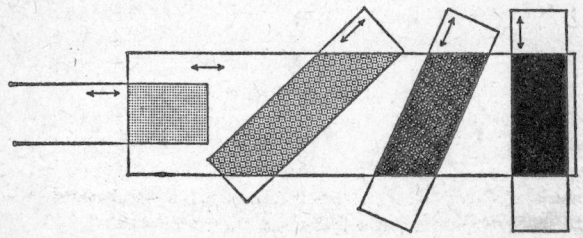

Figure 96. A 'faucet' for light. If two polaroid filters are put one on top of another, they let most light through if the two planes of polarization (double arrows) tally. Less and less light will be let through, the more the two planes diverge.

Does this perhaps polarize the light in the same way as the filter in front of the stovepipe? Then it might act as a sort of filter for light, letting through most light when the plane of polarization of the incident sky light tallies with the lens, and least light as divergence increases.

If you have a pair of polarized sunglasses, you can convince yourself of this. You look at a piece of blue sky through the glasses and turn them. When the sky appears darkest, the polarization directions of light from the sky and the glasses will always be vertical to each other. In this way you can roughly reproduce the graph of Figure 95.

The bees might also do this: only Professor von Frisch's experiments proved beyond doubt that the lenses of the com-

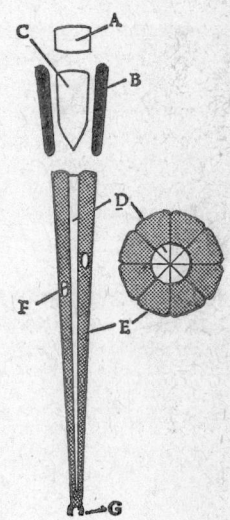

Figure 97. An ommatidium from an insect's compound eye cut lengthwise (left) and across (right). The light falls through the lens (A) and the crystal cone (C), which has a pigment covering (B) that makes each ommatidium proof against light from the next one, into the rhabdom (D), which is surrounded by the visual cells (E) with the nucleus (F). From the tip at the bottom, nerve fibres (G) run to various relay stations.

pound eye (and also the transparent crystal line cones beneath it) do not show the slightest trace of polarization.

So we must pursue still further the course of the rays in a bee's eye. After penetrating the lens, the light falls into a deep narrow shaft. Here nature has anticipated the invention of fibreglass optics: the light is conducted to the bottom of the shaft in eight parallel glasslike optical rods called rhabdoms. These light-conductors are surrounded by a rosette of eight visual cells, equally long but a good deal thicker – joined so that there is a visual cell for each optical rod.

Is this the long-sought device for the detection of polarized light? The decisive discovery was made in 1957 by the Americans, Dr. T. H. Goldsmith and Dr. D. E. Philpott,[11] with the aid of electron microscopy.

If we look at the tiny glasslike rhabdoms enlarged 25,000 to 50,000 times, peculiar structures can be seen. The rhabdom looks like a rack for wine bottles. Innumerable glasslike tubes, each with a diameter of only the 60-millionth part of a millimetre, are horizontally stacked in the rhabdom in such a way that one opening of each of them leads directly into an associated visual cell.

In these countless microtubes lying parallel there is a visual substance sensitive to light, which probably carries out the conversion of light into nerve excitation. Obviously the most powerful effect can be developed by light waves that vibrate in the direction of the tubes.

If we now look at the cross section through a separate eye with the eight rhabdoms, we see that the tubes in these are stacked in different directions.[12] In two adjoining rhabdoms

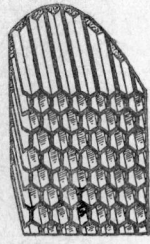

Figure 98. Small section of a rhabdom. The tiny tubes, which become visible only under an electron microscope, are perpendicular to the direction of the light passing through.

they run parallel, in the next two at right angles to these, and so on. From this we can deduce the following: if incident polarized sky light vibrates in the direction of a stack of tubes, the four associated visual cells are excited most, the other four least. So one part of the cells sees "light," the other "dark." This may suffice for a rough recognition of polarization direction; a more subtle definition is evidently then reached by the co-operation of many adjoining separate eyes. Here perhaps is the deeper sense for the mathematically exact order which is so remarkable in enlarged photographs of compound eyes: quite a natural explanation, in fact, of this "supersensory" sense.

Birds' Celestial Navigation

Every autumn great flocks of starlings, from southern Finland, southern Sweden, the Baltic, and Denmark, fly through Holland to winter in the north of France, the south of England, and the south of Ireland. Dr. A. C. Perdeck[13]

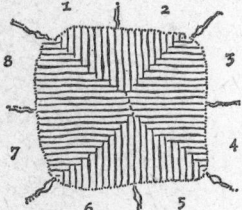

Figure 99. Cross section of the rhabdom in a honey-bee's ommatidium. The tubes come from the eight numbered visual cells around them. In every two adjoining rhabdoms tubes run in the same direction, which is also that of the two rhabdoms opposite.

launched an impressive large-scale experiment with some of these migrants.

In the country around the Hague he caught no less than 11,000 migrants in transit, ringed them, and transported them about four hundred miles south-south-east into Switzerland. They were released between Geneva, Zürich, and Basle. How would they fly now? In the accustomed flight direction south-west, or were they capable of realizing the forced transport and taking counter-measures?

The recapture of many marked starlings gave a surprising answer. All the young birds, which had been hatched during the previous few months in the Baltic area and which had never before in their short lives made the journey, so did not yet "know" their destination, flew south-west; that is, parallel to the course of the members of their species which had escaped the Dutch nets. The information on flight direction was evidently innate, though now, of course, they reached Spain and Portugal, regions hitherto unknown to Baltic starlings.

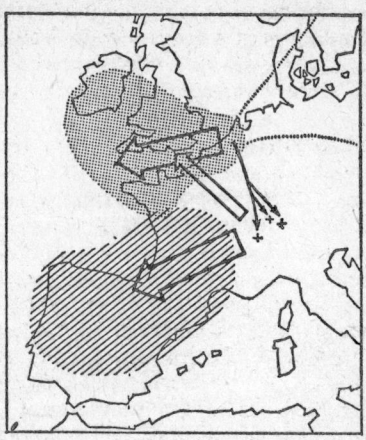

Figure 100. Incident on the journey. Scientists in Holland caught 11,000 starlings making an intermediate landing on their autumn migration from the Baltic area to their winter quarters (dotted area), and took them to Switzerland. After their release inexperienced young starlings carried on in their old migratory direction and settled in the shaded area. Only experienced birds, which had made the journey once or several times before, noticed the deception, and in an amazing feat of navigation altered their course so that they reached their traditional places.

The behaviour of the older birds, with migatory experience in previous years, was quite different. Most of them flew north-west, and on this course, altogether unfamiliar to them, reached their hereditary winter quarters in northern France and southern England.

"So here we are dealing with two different ways of orientation," the zoologist Dr. Klaus Schmidt-Koenig,[14] of Göttingen, comments on these experiments. "Apparently all the young starling takes with him on his first autumn migration is knowledge of the direction he must fly in, and the capacity to maintain that direction through an orientation mechanism. We call this directional ability or compass orientation. The adult bird, on the other hand, possesses the capacity to navigate."

Navigation is the art of making out one's present position and from there following a course which leads to a desired objective. Navigation always has to be used when the objective itself cannot be seen, heard, smelt, or directly perceived in any other way.

Animals and humans can navigate by several fundamentally different methods: terrestrially, by known reference points like lighthouses, sea-marks, churches, trees, roads, coasts, river courses, mountain ranges; celestially, by the position of sun, moon, and stars; also by magnetic fields and forces of inertia.

Before the exploitation of scientific discoveries, however, man cut a very poor figure compared to the animals. Until the thirteenth century, seamen scarcely dared venture out of sight of the coast. The discovery of the magnetic compass was among the things that enabled Columbus to maintain a straight course on the high seas. But if anyone had asked him in mid-Atlantic exactly where he was, Columbus could not have answered. Just like the inexperienced young starlings, he used solely direction or compass-orientation. This is why all the explorers after him landed at different places in the New World, carried by winds and currents.

It was not until 1730, when the mirror sextant was introduced in navigation for determining the exact position of the sun, and a precision ship's clock, the chronometer, was developed, that man was capable of the same achievement as the experienced adult starlings, carried off course by "adverse scientists," who nevertheless succeeded in finding their destination. From then on human seafarers, too, under a clear sky could at any time work out their present position and make port at New York, Cuba, or Rio de Janeiro without having to search up and down the coasts.

Since then men, like migrant birds and insects, have practised both celestial and terrestrial navigation, according to the situation they were in. The captain of a passenger ship normally finds his way by the stars when on the high seas, but by landmarks when in sight of the coast and coming into a harbour; the same goes for birds and bees.

For long stretches of their flight, migrant birds find their way celestially, but they do it by landmarks when they are near a familiar destination. We should not, however, overrate their recognition of the aerial view of home. Many birds that travel extensively, such as the European robin, fly exclusively by night; yet all of a sudden there they are in the same garden in which they were nesting last year. So celestial orientation must have led them astonishingly near to their destination.

On the other hand, it is quite common also for such globe-trotters to correct in flight their celestially adjusted beeline course by landmarks. It seems to be definite, for instance, according to Professor Hans F. von Schweppenburg,[15] that day migrants, which in the spring fly northwards over West Africa, are pushed far off their northern course by the north-eastward bend of the coastline, right until the Rock of Gibraltar comes in sight. Only then do they cross from one continent to the other.

The many American hawks which pass over the Isthmus of Panama twice a year also deviate from the direct course and follow the westerly edge of the Sierra Madre. They probably do this so that they can make use of the range's thermal up-currents for comfortable sailing and gliding flight which will conserve their energy. Such distinctive surface formations of the earth, which influence the bird in its choice of route, which guide it, so to speak, are technically called leading lines.

These include rivers. At the beginning of July, for instance, one can observe gulls near Münden in the province of Hanover, flying down the Werra and then following the Weser farther north; thus they are sure of finding their destination, the southern part of the North Sea. Greylag geese even seem to use "traditional" routes. Leading ornithologists have no doubt that they know exactly every resting-place over the whole distance. The geese evidently pass on their knowledge of these things to succeeding generations, and so the route which experience tells them is least dangerous becomes traditional for them.

The same phenomenon can also be observed with bees. There is soon a regular small air-bridge leading from the hive to a rewarding field of nectar. At first the insects take their beeline by the sun compass. But after repeated flights by the same route they soon deviate from it, steering for conspicuous points to left or right, a tree standing on its own, a fence, a farm. They will then tend to use the known landmark rather than the position of the sun.

But let us return to celestial navigation, which is so much harder than simple compass orientation. For this the creature, an experienced starling, or a homing pigeon, for instance, must know three different things: its present position, the position of its destination, and the geographical relationship of the two points to each other. That means that as well as a compass there must be an ability to determine the map reference. This is the famous map-compass concept of the late Dr. Gustav Kramer,[16] the German zoologist.

Man can find his position this way only with the aid of a precise sextant, an accurate watch, and spherical trigonometry. Even then a practised navigator with all his printed charts and tables will take twenty minutes to find out his ship's position. Yet homing pigeons, a few seconds after being released, will know where they are. How does a bird carry out such position-finding?

So far we have no answer to this. We know plenty of striking facts in the field, but still cannot make a synthesis of them.

Every child knows that an experienced homing pigeon transported to strange territory will very probably find its way back to the pigeon loft. Does the bird memorize the way during transport? To see if this was so, scientists have taken them in darkened cages, driven around mazes of streets, even put them on a rotating turntable, so as to throw them into complete bewilderment. None of it worked. Most pigeons still found their way home; and they really must have started their position-finding navigational operations only on their release.

Things became still more mysterious as a result of a series of experiments carried out by Dr. Kramer. Dr. Hans Löhrl[17] of Radolfzell Ornithological Station gives this account:

aviaries were put up at Wilhemshaven, in which young pigeons with no experience of flight could only fly to and fro a bit. They had no chance of circling to get to know their immediate environment, so they could see no more of the world than the view from the cage allowed. They were not even given the opportunity of seeing their cage from outside.

Then Dr. Kramer took all the pigeons some ninety-five miles south to Osnabrück for their first flight. Many of these novices did more than find roughly the direction of home; they landed right on the roof of the aviary. So they had clearly absorbed their impression of home exclusively from the cage.

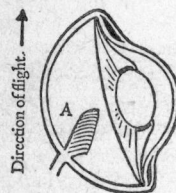

Figure 101. Is the pecten (A) the mysterious sextant in a bird's eye?

The question now was: how much must the pigeon see to grasp the geographical position of home?

In the second experiment Dr. Kramer limited still further the view of other pigeons, equally inexperienced in flight. He surrounded the aviary with a high fence, which allowed the birds no view of the surrounding terrain or the horizon and left only the sky uncovered. These pigeons were also taken to Osnabrück, and none of them found the way home. On release they were completely disorientated. The places where they were found afterwards were scattered in all directions round their starting point.

If pigeons are raised, then, so that they cannot see their surroundings or the horizon, they cannot develop the ability to find their way home. This seems quite clearly established. But it is still a mystery that pigeons should need to see the horizon for celestial navigation.

At present we can suggest only the following hypothesis: the sailor, too, needs the horizon as a reference line when he wants to establish the height of the sun with a sextant. Does the bird also measure the sun's angle of elevation? This would be a great advance on the bees' technique of celestial measurement, for they, as mentioned earlier, can measure only the sun's azimuth, not its altitude.

But here we come to another stumbling block. It is relatively easy to imagine how insects can measure angles through the special structure of the compound eye. But birds have camera eyes like man's, and man's vision is nowhere near good enough to determine the azimuth and altitude of the sun exactly to a degree, let alone to a minute of arc or a few seconds of arc, as would be required for celestial navigation.

Is it not altogether absurd to assume that a bird's eye would be capable of such things? Let us have a close look at a homing pigeon's eye in cross section.

The fundamental difference from our eye is obvious. A comb-like structure sticks out of the retina into the optical interior. It lies over the place where the optic nerve enters the retina, the place, in fact, where men have their "blind spot." This strange structure is called the pecten.

Its purpose was for a long time completely mysterious. Scientists even thought of it as one of nature's mistakes, since this "palm-frond" would surely hinder projection of images onto the retina. We still do not know the organ's exact purpose. We may presume, perhaps, that this structure with its strong blood circulation is a component of the "bird sextant." Perhaps the shadow it casts onto the retina is probed by the visual cells and is used by successive nerve circuits as a protractor for measuring the sun's position. But perhaps it all works quite differently. Whatever the answer, there can be no denying that a bird is capable of measuring angles considerably more exactly than man can do unaided.

In the confident belief that homing pigeons can really make such exact angular measurements with the pecten or in some other way, Dr. G. V. T. Matthews[18] of Cambridge University and Dr. C. J. Pennycuick[19] of the University of Bristol have suggested hypotheses along similar lines.

Dr. Pennycuick imagines things in this way. When a bird has been carried away somewhere and is starting for home, it measures within seconds the angle of the sun's height over the horizon. At the same time it "remembers" how high the sun was at the same time of day at its home. If the sun is now higher, that means the bird has been taken south and so must fly north to get home. It could notice, similarly, whether it had been taken east or west. This would work, at least in theory, by a comparison of the speeds with which the sun changes its height.

Here is a simple example. Let us assume that, seen from the pigeon loft, the sun is in the south at exactly twelve noon. Then it will not change its height for a minute, for it is just reaching its peak. If the pigeon is taken west, the sun will not have reached its peak by twelve noon (home time). It is still climbing, and does so the faster, the farther west the pigeon has been taken (see Figure 93 on p. 266). The pigeon could decide from this that it must keep going east.

It would also have to reach a synthesis from the north-south and east-west components in order to find out the direct course home. Isn't that asking a bit much of the bird? Possibly, yet how else should it navigate?

At any rate the pigeon would have to possess the following abilities: a time sense, or rather an inner chronometer, that always indicates with unshakable precision what the time will be at its home; the capacity of measuring passing time exact to seconds, and of registering changes of the sun's position of the order of magnitude of minutes of arc. The bird would also need practically to see the sun's course as movement, as we do when we look at it through a stationary telescope.

Unfortunately it is tremendously difficult to verify this experimentally. So far it has not been possible either to prove or refute the theory. In particular, there are some strange things which are not explicable in terms of it. The theory would lead one to assume, for instance, that when a pigeon is starting home, it would take a more exact course, the farther away it has been taken, for it would be able to recognize long distances it has been taken better than short ones. But that is not at all the case.

Up to a distance of about sixteen miles from home, pigeons as a rule fly very accurately towards their destination, according to Dr. Klaus Schmidt-Koenig.[20] Between sixteen and seventy-five miles, however, there is a "dead zone," from which the birds either do not find their way home or do so with difficulty. Beyond this distance they again fly pretty exactly in a homeward direction.

Does this mean that we have to reckon with two different methods of celestial navigation, one for shorter and one for greater distances? But why does the pigeon's navigation sense in the "dead zone" receive no or inadequate information? One question after another – and no answers.

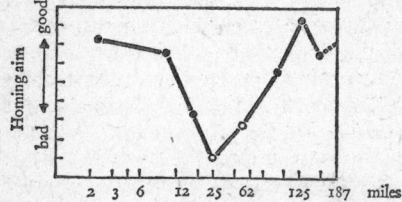

Figure 102. From small and great distances homing pigeons find their way to the loft equally well. Between 16 and 75 miles away from home there is a 'dead zone,' however, which so far no one has been able to explain.

Of other baffling things I will mention only one: breeders of homing pigeons on Hawaii arrange their homing competitions only at night; and *their* birds, too, find their way home magnificently – without sun! The European pigeons also are not completely lost if during their flight the sun is hidden behind a thick cloud. Their homing performances are worse then, but many of them do reach home. What do they steer by when there is no sun for them? By the terrestrial magnetic field? In the last section of the book I shall examine this question more closely.

The navigational problems of the homing pigeon (a descendant, incidentally, of a sedentary bird, the rock-dove) appear relatively simple if we think of the long-distance record-holders among migrant birds like the albatross and the shearwater (see pp. 242–4). But clearly there is a long way to go, and much research still to be done, before the problems of animal navigation are solved.

Celestial Navigation by Night

On 24 October 1963 the inhabitants of the town of Münster had an eerie experience. The traditional autumn fair had just begun that evening, radiating a bright dome of light from thousands of lamps into the cloudy sky above the West-phalian capital.

Suddenly, shortly after seven o'clock, the air was filled by an excited, cawing clamour. It sounded like hundreds of voices crying for help. Three local zoologists[21] recognized overhead a huge wedge formation of three or four hundred cranes. The high clouds had become thicker and thicker, thus robbing the birds, which fly by night as well as day, of their normal orientation. In their distress they were evidently keeping to the sea of lights from the earth.

An effect which never occurs in steering a course by the position of the infinitely distant stars was now produced by the star-substitutes, the city's lights. The cranes were trapped by these lights of the fair, and for five and a half hours circled continually in a clockwise direction over the centre of the city. Each "round" took them from fifteen to twenty-five minutes.

At half past twelve, when the lights from the fair and most of the neon lighting in the rest of the city had gone out, and at the same time the high cloud formation lightened a bit, the crane squadron at last flew off in a south-westerly direc-tion.

Cranes probably steer by the sun in the day and the stars at night. It must have disturbed them intensely when the sky was enveloped in cloud and mist, and at the same time un-familiar constellations appeared deep below them. But for the city illuminations they would presumably have landed and had a rest until celestial navigation was again possible for them. As it was, however, the conflict of instincts obviously left them completely disorientated and literally in the air.

If somebody had declared in 1955 that birds flying at night navigate by the stars, he would certainly have been considered crazy. Today there is little doubt among scientists that animals do possess this amazing faculty.

Till then there seemed to be an insurmountable obstacle facing research on the nocturnal flight of small song-birds: how could anyone ever follow which direction individual migrants took in the darkness (as a rule they do not fly in formations) and how they behaved on the way?

In 1948 Dr. Gustav Kramer had already made a crucial observation when he discovered the use of the sun-compass by starlings (the same year, by the way, in which von Frisch established its use by bees). It struck Kramer that right from their take-off migrant birds fly exactly in the right direction. So they do not take off into the wind like an aircraft, nor do they fly up in any direction and then circle a few times to get their bearings as homing pigeons sometimes do. While still on the ground they know their direction of flight.

This is a great advantage for the researcher. He does not need to follow the birds by radar or by miniature transmitters tied under their bellies. He need only put the bird in a special cage, watch it fluttering about on a ring perch until it decides the correct position for take-off and indicates by a flurry of wings that if released from the cage it would now take off.

If the cage is now turned slightly, the bird at once adjusts its position, first moving a little too far in the required direction, then nervously and tensely swinging back to the correct position, behaving in fact just like the magnetic needle in a compass. Birds behave in basically the same fashion

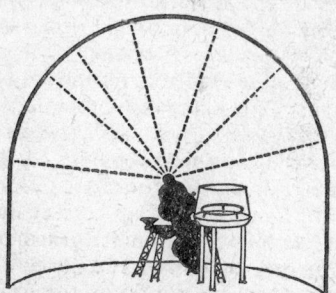

Figure 103. The interior of a planetarium. The projector (black) throws star patterns on the dome with fine light rays (dotted lines). Near the projector is the blackcap cage with the circular perch.

in actual migration. They continually diverge from their true course by about five degrees to right or left and then correct it.

As the bird gives perfectly reliable information about its flight direction in its cage, there was no reason why experiments should not be carried out in a planetarium.

It was two German zoologists, Dr. Franz Sauer[22] and his wife, who carried out a series of planetarium experiments which have become world-famous. The idea came to Dr. Sauer in 1956 when he noticed that, when the sky was clouded, blackcaps and other European (*Sylviid*) warblers would orient themselves for migration only if a few bright stars were visible. That provided him with the long-sought clue: night migrants must navigate by the stars.

The experiments were carried out in the Olbers Planetarium of the Bremen Maritime Academy, and they had their snags. Dr. Sauer and his wife took it in turns to lie on their backs below an experimental cage in the glimmer of the dim artificial starlight of the dome-shaped hall, watching the birds take their bearings and noting these down. This might have been tolerable for a few days, one would think; the Sauers kept it up during the migatory season twice a year for several years.

Also, the blackcaps got very cross when they were fetched from their comfortable cage to be experimented with. Birds which Dr. Sauer had tamed had to be prevented from recognizing him when he experimented with them; otherwise they grew hostile and shunned him for weeks on end. Even returning them to their homes and tempting them with freshly skinned mealworm larvae failed to restore relations to the old basis. The offended birds refused to accept these dainties from him, though taking them eagerly enough when offered by his wife. So he always had to put on a mask and disguise himself like a bank robber for the experiments.

He began by showing his blackcaps at the autumn migration season an artificial starlit sky which looked exactly like the real sky over Bremen. They were completely deceived by it, and fluttered without hesitation in the direction of Turkey, to the south-east.

That, of course, proved nothing at all. The birds might not have been taking their bearings from the artificial stars of the planetarium, but from something quite different, such as the earth's magnetic field, perhaps, or electro-magnetic rays proceeding from space.

To meet these objections, Dr. Sauer turned the projected image of the sky in all directions; but whatever he did, the birds always took off in the direction which should have been south-east by the position of the stars, even if this was really north or west, for instance.

Moreover, if in the course of the night the stars in the planetarium did not travel across the sky like real stars, but remained stationary for several hours, the birds adjusted their flight direction accordingly. As with the ants (see p. 267), they seemed to know that at 11.00 p.m. in September this or that constellation would be in the south-east, and thus in their flight direction, but that it moved on, and by 2.00 a.m., say, would be in the south. They seemed to know, too, that at that time their course must not be exactly in the constellation but 45° to the left of it.

These experiments made it clear that blackcaps instinctively recognized individual constellations, even when there were only isolated stars visible through gaps in a cloudy sky. But if the whole sky is overcast or the whole dome of the planetarium is in darkness, they flutter helplessly about the cage for some time before deciding that they cannot hope to find their bearings on this occasion and dropping off to sleep. In natural conditions they simply interrupt their migration in these circumstances.

How do blackcaps acquire their extraordinary astronomical capacity? Do they learn it from their parents? To test this possibility, Dr. Sauer kept a blackcap in a closed room from the time it was hatched. For several months it never saw another blackcap, or the sky, either by day or night.

Such creatures growing up in complete solitude and isolation from their natural environment are known to German scientists as "Kaspar Hauser" specimens, after the mysterious foundling of 1828, who was kept in a locked room until his sixteenth year, had not previously seen a human being, animal, or plant, did not know the difference between night and day, could not speak, and did not know whether to walk upright or move on all fours. Kaspar Hauser animals are

reared for the purpose of differentiating acquired and inherited modes of behaviour.

In September, when this Kaspar Hauser blackcap was
seized with migratory restlessness, Dr. Sauer put it in the
Bremen planetarium and switched on the stars. At first it
was thoroughly startled, but then it flew off towards the
south-east!

So blackcaps must have inherited their knowledge of
celestial geography and the course of the stars. There can
hardly be any other interpretation of the facts, although we
have not the slightest inkling how nature can have produced
anything so fantastic. There seems to be something similar,
however, in the deep-sea world, where members of a fish or
octopus species recognise each other by the "constellation"
of their light-organs.

Dr. Sauer now took matters a stage further. He deluded his
blackcaps into believing that they were really flying south-east
on their natural course. He caused new constellations to
appear in the south, while familiar ones dipped beneath the
northern horizon. After showing the blackcaps the night sky
as it was over Bremen, he showed them in succession that of
Prague, Budapest, Sofia and western Turkey. Still they headed
due south-east.

But then something remarkable happened. When he showed
them the night sky of the eastern Mediterranean, as it is to
be seen in the sea area around Cyprus and in Israel, they
suddenly turned south, like a ship changing course. This
celestial turning-point saves them from going on into the
Arabian desert, where they would certainly die a miserable
death. Instead, they are safely guided along the Nile towards
their winter quarters.

It was very striking that in every case the bird at once
turned south on seeing the starry sky as it would have been
over Cyprus, irrespective of whether it had been deluded into
the belief that it had flown from Bremen to Cyprus in three
weeks or in the "record time" of three hours. This shows that
the blackcap can not only keep to a compass course by the
stars but can also recognize its geographical position; that is,
navigate.

Its destination too is written in the stars. If in the planetar-

Figure 104. The two directions of the flight of the blackcaps (Sylvia curruca) are from Germany to Africa. The turning-point is in the area of Cyprus – Israel.

ium it is tricked into believing that it has reached the area south of the great loop of the Nile, it feels it has arrived. Its nightly migration fever disappears, and it goes to sleep.

Other migratory birds are led irresistibly by the stars to other destinations. The uncanny impulse is far stronger than the desire to eat, and outweighs the attractions of a comfortable life in any of the regions the birds pass through on their route, no matter how favourable the climate or abundant the food these may have to offer.

The western European whitethroats and garden warblers, which migrate to South Africa by way of Gibraltar and West Africa, would find ideal living conditions for instance, on the Okawanga River in Angola. But after only a short rest they abandon this fertile region and fly 160 miles across the northern part of the Kalahari desert to the Etosha Pan and into the arid border areas of the Namib desert, where they have a hard struggle for survival.

The only explanation of this behaviour is that the drive to reach a destination indicated by a particular position of the stars is stronger even than the drive to eat and drink. In their egg the young blackcaps have been given a travel ticket for life, as it were, and they cannot escape the obligations that go with it.

Dr. Sauer, who is now working at the University of Florida, hopes to find out which of the many stars the night migrants use to guide them. The simplest hypotheses were that they use either the Pole star, which does not move, or the bright ribbon of the Milky Way, but both have been disproved. It is a simple matter in a planetarium to darken an individual star or cut out the Milky Way, but doing so made no difference to the birds.

Bright moonlight, shooting stars, and summer lightning confuse them. When shown these natural phenomena in their special cage out of doors at night, they were startled for a moment, broke off their south-easterly whirring, and for a short time flew in the direction of the startling phenomenon.

It must be pointed out, of course, that they did this only in Dr. Sauer's cage, which was so constructed that they could see only the sky and no landmarks on the ground or the horizon. If they are allowed ground visibility such as they have on their night flights, in the case of "celestial disturbances," they promptly switch over to using landmarks and navigate by these for a time. On long nights of full moon they interrupt their journey. But with a sea of lights from a big city the same thing could happen to other migrant birds as happened to the cranes over Münster.

Strangely enough, night migrants are not in the least bit disconcerted by the appearance of the bright planets Venus, Mars, Jupiter, and Saturn. So the investigator is left with the difficult and tedious task of finding out by trial and error in the planetarium which of the constellations guide the birds over strange lands by night.

The question which constellation in the glittering fullness of the clear night sky actually constituted the birds' signpost could be answered only in 1967, by Professor Stephen T. Emlen, now at Cornell University.[23] The bird used was the indigo bunting, a small songbird that migrates annually from the southern United States from 1,800 to 2,500 miles to its South American winter quarters, and returns in the spring.

This investigator originally tried to show, at the planetarium of the University of Michigan, similarly to Dr. Sauer, that the Milky Way or the Big Dipper served as the guidepost. But he tried in vain. Experiments with the constellation Cassiopeia or with Polaris alone were also unsuccessful. But when Professor Emlen eliminated all stars in the artificial sky that were contained within 35° of Polaris, his birds were immediately disorientated and ceased to flutter.

Professor Emlen believes therefore that he has found the guidepost in the totality of stars that belong to the Big and Little Dipper, Cassiopeia and the constellation Draco. Different birds, it appeared, are guided individually by somewhat varying star patterns. A few stars on the periphery may be absent, but the loss is made up by other stars that are meaningless to the bird's mates. In any case, indigo buntings must, when they migrate south in autumn, look back over their shoulders towards the north, in order to keep to their correct course.

To leave birds for the moment and come to insects, we already know some interesting things about ants. They are nonstop workers, doing three shifts a day; they can sleep enough during the winter. Now and then one of them may doze off for a bit; but if in doing so it gets in the way of the other workers, they jostle it, which startles it into carrying on.

So there is a lively traffic day and night on the ant trails in the woods. Scouts that turn off these trails find their bearings in the daytime by the sun's position and at night by the moon's. They are very well able to distinguish between sun and moon, which pass across the sky in paths of very different steepness, and so are very different from each other in their azimuth curves (see Figure 94, p. 267). The ants take these differences fully into account when working out their course. They evidently have two separate systems for direction orientation, a sun compass and a moon compass.

What about the stars? Can insects, like night-migrant birds, recognize stars and perhaps also take bearings by them? Scientists still disagree on this, but from the experiments of the Berlin entomologist Dr. Karl Cleve[24] it seems possible that moths too can register the light of the stars.

Here again a discovery was due to chance, as has happened so often in the history of science. Dr Cleve is a passionate lepidopterist, and has already enticed thousands of moths with lamps. In the process he tried out a great many different types of lamp with various spectra of light, simply to find out which produced the most efficient relation between electricity consumption and amount of insects arriving. It turned out as a result that the greatest sensitivity to light in a moth's eye lies in a colour range of a shorter wavelength than that of the human eye: in violet (not ultra-violet) with a wavelength of 400 to 425 nanometres.

This "violet displacement" must have a reason, Dr. Cleve decided. So he had the idea of comparing the curve of sensitivity to spectral light in the moth's compound eyes with the curve of spectral light intensity in the stars. This is rather different with each type of star. But if we take the average of all the 3,000 stars which we can see in our latitude with the naked eye, there is an amazing correspondence between it and the sensitivity of the moth. Qualitatively, therefore, the moth's eye is ideally adapted to the light of the stars, while the human eye is more suited to daylight.

But is starlight also bright enough for moths to be able to recognize it? To settle the question, Dr. Cleve carried out a series of experiments with lamps of different intensity and compared the results to the intensity of light from the stars. His findings were that moths can register almost as many stars as man can.

We do not yet know for certain *why* they should take notice of starlight. Possibly they steer in the darkness directly by moonlight or starlight, because they can then be sure of not colliding with an obstacle. This hypothesis is supported by the observation that lamps have more effect in country where visibility is limited than in open country.

It is also possible, however, that starlight allows moths a certain stabilization of course on scent-directed flights to their females or food; or, in the case of females, to their egg-laying places; or even on distant flights. But on these matters at present we are still deep in pure speculation.

The Magnetic Sense

Magnetic senses can no longer be dismissed as humbug or occultist fantasies. They exist as a zoological reality. The most impressive proof of this was provided in 1965 by Professor Friedrich Wilhelm Merkel[25, 26] of Frankfurt University.

The history of this rather sensational discovery started in 1957 by chance. In the Zoological Institute laboratory, Dr. Hans Georg Fromme,[27] a colleague of Professor Merkel's, kept several robins. At night, during the autumn migration period, the birds began to flutter about restlessly in their cage. Animal lovers are familiar with this instinctive outbreak of the urge to travel. But the Frankfurt zoologist observed something more than that: the robins (which migrate at night) all tried to fly up only in a quite definite direction, south-westwards. This is exactly the course they must follow on the great flight which takes them from Germany to Spain. Unlike Dr. Sauer's blackcaps, they could see from their cage neither the natural sky nor the artificial one of the planetarium. They were within the walls of the Institute, and the shutters were tightly closed. Evidently they must have felt something through the walls of the building which indicated to them the right "point of the compass."

The experimenters then put the robins in a steel chamber; and here they behaved quite differently. Full of migratory restlessness, they still fluttered up to the bars of the recording cage, but without showing the slightest tendency towards a particular direction. When the steel chamber's armoured door was opened, they again immediately showed a slight inclination to flutter in the direction of Spain. So the steel chamber must have been capable of keeping the robins from their mysterious signpost. As the lines of a magnetic field are weakened by a steel chamber, these findings forced the scientists to consider whether the magnetic sense so long tabooed might not after all exist.

It still had to be proved, however. To make things easier, I shall describe only the experiments which eventually led to a successful conclusion. In fact it took several years full of setbacks and doubts.

Tests had repeatedly been made, by other researchers as well as on other birds (such as homing pigeons), in which small bar magnets were hung around the birds' necks or fixed to their wings. It was thought they would be bound to be troubled by these if they really did have a magnetic sense. But as the magnets did not disturb their capacity for orientation in the slightest, the results were taken as proof that there *was* no magnetic sense.

A similar thing happened in tests with large homogeneous magnetic fields, their direction turned against that of the field of terrestrial magnetic force. For instance, two magnetic coils, each about six feet in diameter, were set up inside steel chambers, and the birds simply put between these coils. But they made no move to find their bearings by the artificial field. Once more the existence of a magnetic sense seemed to be refuted.

The deeper cause of these failures lay in a false conception of the magnetic sense postulated. For we should not imagine a sort of magnetic needle which in seconds adjusts itself to any magnetic field.

Things are far subtler than that. Above all, the birds must become used to altered magnetic relations for quite a while, before they will react to them. The strength of an artificial magnetic field by which they are to be influenced must correspond exactly to the natural terrestrial field. If it is made even slightly stronger or weaker, the birds evidently cannot register it any more or it loses its meaning for them, unless they are given the chance to get used to the new strength for several days.

In Frankfurt the total strength of the field of terrestrial magnetic force is 0·41 Gauss. Within the steel chambers described above it was weakened to 0·14 Gauss. The robins therefore at first were completely disoriented. But when Professor Merkel left them in the room permanently for several days, they very gradually began to resume their tendency to flutter towards the south-west.

Now that the need to acclimatize was taken into account, the experiments with fields of the large magnetic coils also brought the hoped-for results. When the experimenters turned the artificially produced magnetic field against the sense of the natural field strongly suppressed in the steel chamber, the robins were diverted in their take-off direction just like Dr. Sauer's blackcaps under the distorted artificial firmament of the planetarium.

"In our judgment," writes Professor Merkel, "the experimental results to hand are clear evidence that in finding their direction during their migatory restlessness the birds use the field of terrestrial magnetic force. So far the results of current experiments confirm us in this view."

Accordingly we have the following picture. On an autumn night the robin is seized by migratory restlessness. It looks towards the stars, recognizes from them the south-westerly direction which will lead it to Spain, its distant destination, and flies off, quite alone, without its mate, young, or any companions – though other robins, of course, do the same quite independently. But as soon as the night sky is overcast and the robin can no longer recognize any signpost stars, it does not, like the blackcap, need to land and wait for clear weather again. Like a sailor on the high seas, it switches over to steering by magnetic compass and flies on unerringly over rivers, lakes, and mountains, till it has reached its destination in that distant country.

Might the discovery of the magnetic sense also shed a small gleam of light on the mysterious homing powers of many kinds of animals?

It has already been mentioned that homing pigeons can find their way home even at night in a completely overcast sky. Weddell seals[28] dive under the fields of dense pack-ice in the Antarctic down to a depth of 1,150 feet and for a distance of twenty miles without surfacing. Despite the darkness under the pack-ice, they follow a straight course to their usual ice-free haunts on the coasts of Antarctica.

Wild mice[29] have been trained to follow particular compass directions when optical, acoustic, and chemical landmarks were excluded. Many other tests have indicated that laughing gulls and certain other birds, and also snails, can find their way home without being able to see, hear, or smell the direction of home.

So senses beyond our grasp, which sound quite uncanny, must be more widespread in the animal world than we once believed. This does not even mean that all the creatures just mentioned must necessarily find their way magnetically; that has not yet been proved with them. But a magnetic sense *has* been proved for the following as well as robins: termites,

June bugs, pond snails, weevils, crickets, locusts, wasps, and and flies.[30]

Here too chance played a crucial part. At the end of 1963 Professor Günther Becker,[31, 32] of the Federal Institute for the Testing of Materials in Berlin, had received a sample of young termite queens from Rhodesia (*Macrotermes* and *Odontotermes* species). In the evening he quickly shook them onto the bottom of a breeding box, where they lay in all directions. The next morning he saw to his astonishment that they had all settled down to sleep lying exactly in an east-west position.

He cautiously turned the box by 90°, but after a few hours the insects had corrected their position so that their heads were facing either directly east or west or at right angles to an east-west line. Like inert magnetic needles they had eventually found their correct position.

Like the robins, however, the termites could no longer find their direction when put in a steel box with thick walls. But when Professor Becker fixed a bar magnet above them within the box, a gradual movement again went through the termite assembly, and between a quarter of an hour and several hours later they were all lying directly across the axis of the magnet.

Why they sometimes reacted rapidly and at other times much more slowly, we do not yet know, nor do we know about the directions of orientation by the magnetic field. True, we can now understand how it comes about that the queens of many termite species are always found, when their hills are laid open, in an east-west or north-south position; but we have no idea at all of why they *have to* be aligned in these directions.

The only exception in the matter of direction is the Australian compass termite. This species builds fortresses up to thirteen feet high and ten feet long, but only three feet wide. These "walls" rising out of the ground all run from north to south. This way of building has the great advantage for the termites that the sun shines into them with almost the same warmth throughout the day: in the mornings and evenings it irradiates the whole length, but at midday only the narrow width.

A function for magnetic and electric fields was discovered in June bugs (cockchafers) by Dr. F. Schneider[33] at the Swiss Federal Experimental Institute at Wädenswil near Zürich. He dug up June bugs still resting in the ground, kept them for some time rigid at freezing point, then slowly thawed them out at 68° F. (20° C.). They immediately began to roam around, but after a while settled down to rest again. Their resting position, however, is by no means arbitrary; it is determined by invisible forces, which can penetrate even through thick stone walls.

In difficult experiments Dr. Schneider discovered that the June bug is controlled by electrical as well as magnetic fields. Any angle by which artificially produced lines of magnetic and electrical force are shifted in relation to each other causes a quite specific period of unrest in the insect and an equally characteristic direction of the resting position. The direction may often be the mean between the two types of lines of force.

But considerable variations sometimes occur, which is very peculiar. On some days, in fact, nothing would work at all for Dr. Schneider. Regardless of the direction of the artificial magnetic and electrical lines of force, the June bugs would lie down to rest in the geographical north-west/south-east direction or straight across it.

Many people assert that they can sleep well only when their beds are facing a particular geographical direction. Whether this feeling has something to do with terrestrial magnetism or is only imagination has not yet been investigated. Nor can we say at present whether a similar feeling leads the June bug to its remarkable behaviour. But it is established that the magnetic compass can be very useful to the insect in orientation.

On June bugs' orientation the French investigator Roberts, cited by the zoologist Otto Koehler[34] of Freiburg, reports some remarkable things. As soon as the insect is hatched, it leaves on a foraging flight. After a few circles and zigzags the females fly in a straight line to the outlines of a wood nearby, often the one highest on the horizon. The males after a few detours follow the same course. When a smoke-screen was put in front of the wood diagonally across the flight direction, many of the insects flew along it and round the end of it direct to the wood. Others flew over the smoke-screen.

When females while feeding have become ready to lay their eggs, they at once, without first circling to get their bearings, fly back in the opposite direction, straight to the field where they were hatched themselves, and lay their eggs there.

Females caught on their flight to the food, whose eggs had matured in the cage, were transported to a distant region. Without any searching, they also, in unknown country, chose the same direction for the return flight to the presumptive birthplace as the members of their species which had not been caught. After laying their eggs, June bug females fly towards the wood in the same direction as twenty or thirty days before on their first flight.

This direction is evidently imprinted during that flight; and to keep to it, they must see the sky. The sky may be in cloud, and even at dusk they find the way. Evidence is still lacking for the manner of their flight orientation. According to Dr. Schneider's findings, however, orientation by a magnetic field is at least in the realm of possibilities.

A further curiosity was discovered by the American zoologists, Drs. Frank A. Brown, Jr., M. F. Bennett, and H. M. Webb[35] of Northwestern University, Evanston, Illinois, with the pond snail (*Nassarius obsoleta*), which lives on the bottom of inland lakes and pools. These molluscs always steer a quite definite course to the magnetic North Pole; but the angle changes during the day as with the dung-beetle's course by the sun.

With the snail, however, all this is far more complex. We can perhaps come nearest to the facts by a theoretical experiment. Let us assume the snail has a compass needle in its body, which is tied to the minute hand of a watch and which with the minute hand makes one revolution a day. The snail feels compelled to steer in such a direction that the magnetic needle on the "hand of the inner clock" is always pointing north. If this were the only orientation mechanism, the snail would be slowly crawling round in a circle.

But there is a second mechanism. Besides the compass needle which turns according to the sun's course, the snail contains another compass needle within itself which turns by the moon's course. Between the two directions in which the two compass needles would steer it, it chooses a compromise.

With the homing pigeon the complications are still greater, and here for the moment everything is in the realm of hypothesis and speculation. Still, a brief glance at this should not do any harm. As Sir Julian Huxley once said, scientific research is made just as unprofitable by too little imagination as by too much speculative adventurousness.

Research into the homing pigeon's magnetic sense by German scientists stopped short in 1933. In 1965 a link was made with the point they had reached, by a comparative outsider to the field, a mathematician, Lester Talkington,[36] then at IBM's Systems Research Institute in New York. He claimed that the pigeon has in its eye a small magnetic compass, and that it is this which shows the bird the direction of home when the sky is overcast and it can no longer take bearings by the sun's position.

By 1933 the Berlin geomagneticist Professor Hermann Reich[37] had carried out informative experiments with homing pigeons belonging to the German army. He took the birds from Berlin to near the Kyffhäuser Mountains in North Thuringia, where owing to underground deposits of iron the lines of force of terrestrial magnetism are not normal.

Normally things are as follows: if we consider also the vertical inclination of the lines of force, we notice that those at the magnetic North and South Poles are exactly perpendicular to the earth's surface, while at the Equator they run parallel to it. So we can say that between magnetic Equator and North Pole, the farther north a point is on the terrestrial globe, the more steep as a rule will be the angle of inclination of the lines of force to the earth's surface. But in the Kyffhäuser Mountains there is a reversal, and Professor Reich's pigeons promptly took off south, the wrong direction, instead of north.

Lester Talkington has come to something like the same results on the basis of theoretical reflections: the geographical latitude of each point on the globe can be established (apart from certain deviations) by the steepness of inclination of the magnetic lines of force. A pigeon taken to strange country which is capable of recognizing this steepness must always know whether it is north or south of its loft.

This gives a first clue to the way the terrestrial magnetic field can be used not only for compass orientation but also for navigation. It is a good deal harder, though, to say whether an animal can recognize from the terrestrial magnetic field that it has been transported from east to west. At present, no doubt, the question cannot be conclusively answered, but Talkington has a hypothesis ready for this too. He uses the angle of declination, the angle measuring the deviation of magnetic north from geographic north, for this purpose.

Talkington compares orientation by means of the magnetic field with orientation in a "field of sun rays." It is as if animals with a magnetic sense registered in the north a kind of "magnetic sun," except that this always stays in the same place, so the time of day need not be taken into account when working out the angle of the course. In this respect, therefore, the magnetic compass would be even simpler than the sun compass.

With forager bees and ants we already know (see p. 53) that they sum up vectorially all angles and distances even on mazelike zigzag routes, and so can find their way home again from anywhere. Do homing pigeons perhaps bring off the same feat with angles of their course, which they steer in relation to magnetic north?

To prove his hypothesis, Talkington reanalysed some older data obtained by Harold B. Hitchcock. The latter's loft was at Fort Monmouth, New Jersey. From here he took the birds some ninety miles west to New Hanover, Pennsylvania, with a special ulterior motive: as lengthy measurements of terrestrial magnetism had shown, there were two geographically quite different routes from New Hanover to Fort Monmouth with almost equivalent magnetic characteristics; so if the pigeons had a magnetic sense, it would lead them homewards on both routes. How would they fly?

After releasing them, Hitchcock followed them in a private aircraft and kept them continuously under observation through binoculars. The birds did use both routes, although one of these made a huge detour almost to Philadelphia.

In a second test he took the same pigeons nearly a hundred

miles north, to Wurtsboro, New York. With his highly sensitive instruments he had here discovered a place from which a blind alley of terrestrial magnetism led to somewhere quite different from Fort Monmouth, though with the same magnetic characteristics as that home town for the pigeons. Again the theory was confirmed: the birds flew into the blind alley and circled around aimlessly all day in the magnetically right but geographically wrong place.

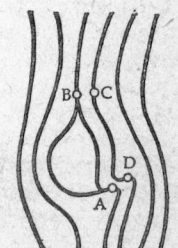

Figure 105. Do homing pigeons take their bearings by the terrestrial magnetic field? The lines connect places with the same magnetic characteristics. According to Talkington, if homing pigeons are taken from A to B, they return home by two different routes, as there are two routes with the same magnetic characteristics. But if the pigeons are taken from A to C, they do not find their way home and roam around for days in D.

Talkington already has an idea, too, as to where the pigeons' magnetic compass might be situated: in the pecten (see Figure 101, p. 280). Further research is needed, however, to obtain more precise details of how a magnetic sense works.

So in the immense field of research on the senses every new discovery raises new questions and hypotheses, which are of absorbing interest even outside the scientific world. Today we have already gained penetrating insights into the almost "science-fiction" techniques whereby nature produces contact between her creatures and their environment. Yet man's spirit of discovery is still like that of an astronomer faced by an infinity of hidden wonders. Clearly it is tremendously difficult to recognize the processes through which we recognize the world.

It may sound sad to many people that our picture of nature is now being more and more resolved into a kind of vast industrial area full of mechanisms that can be technologically described. On the other hand, this development brings with it an invaluable advantage: biology can make the step, from a science confined to nature, to technological application. The biologist can report to engineers on technical inventions in nature which they had till now considered Utopian. This will give him increasing entry, as it is already doing in America, to the laboratories of industry, where he can offer decisive impetus to the advancement of our civilization.

Source Notes

Chapter I

1. ANTON HAJOS, "Die optischen Fehler des Auges," *Umschau*, 64 (1964), pp. 491–6.
2. IVO KOHLER, "Experiments with Goggles," *Scientific American*, Vol. 206, No. 6 (May 1962), pp. 63–72.
3. DEREK H. FENDER, "Control Mechanisms of the Eye," *Scientific American*, Vol. 211, No. 1 (July 1964), pp. 24–33.
4. NORMAN CARR, *Return to the Wilderness*. E. P. Dutton & Co., Inc., New York, 1962; William Collins Sons & Co. Ltd., London, 1962, pp. 110–11.
5. LORUS and MARGERY MILNE, *The Senses of Animals and Men*. Atheneum, New York, p. 226.
6. ROY M. PRITCHARD, "Stabilized Images on the Retina," *Scientific American*, Vol. 204, No. 6 (June 1961), pp. 72–8.
7. JEROME Y. LETTVIN, "Two Remarks on the Visual System of the Frog," *Sensory Communications*, Cambridge, Mass., M.I.T. Press, 1961.
8. R. W. SPERRY reports on preliminary evidence bearing on this subject in his article, "The Growth of Nerve Circuits," *Scientific American*, Vol. 201, No. 5 (Nov. 1959), pp. 68–75.
9. For further details see SIR JOHN ECCLES, "The Synapse," *Scientific American*, Vol. 212, No. 1 (Jan. 1965), pp. 56–66.
10. STEPHEN W. KUFFLER, "Discharge Patterns and Functional Organisation of Mammalian Retina," *Journal of Neurophysiology*, Vol. 16, No. 1 (Jan. 1953), pp. 37–68.
11. DAVID H. HUBEL, "The Visual Cortex of the Brain," *Scientific American*, Vol. 209, No. 5 (Nov. 1963), pp. 54–62.
12. LETTVIN, MATURANA, McCULLOCH, and PITTS, "What the Frog's Eye Tells the Frog's Brain," *Proceedings of the Institute of Radio Engineers*, 1959, No. 47, pp. 1940–1951.
13. LETTVIN, MATURANA, McCULLOCH, and PITTS, "Anatomy and Physiology of Vision in the Frog," *Journal of General Physiology*, 1960, 43 Suppl., pp. 129–75.
14. HEINZ VON FOERSTER, "Biological Ideas for the Engineer," *New Scientist*, Vol. 15, No. 299 (9 Aug. 1962), pp. 306–9.
15. M. B. HERSCHER and T. P. KELLEY, "Functional Electronic Model of the Frog Retina," *Bionics Symposium 1963*, Contri-

buted Paper Pre-Prints, Wright-Patterson Air Force Base, Ohio.

16. WILLIAM BEEBE, *Half Mile Down*. Duell, Sloan, and Pearce, New York, 1951, pp. 124, 127, 168–70, 207, 212, 222–4.

17. EARL S. HERALD and DEITER VOGT, *Living Fishes of the World*. Doubleday & Co., Inc., Garden City, N.Y., 1961, p. 106.

18. N. B. MARSHALL, *Tiefseebiologie*. Verlag Gustav Fischer, Jena, 1957, p. 239.

19. *Ibid*. p. 263.

20. LORUS J. MILNE and FRITZ BOLLE, *Knaurs Tierreich in Farben – Niedere Tiere*. Droemersche Verlagsanstalt, Munich, 1960, p. 281.

21. HANS-ECKHARD GRUNER, *Leuchtende Tiere*. Neue Brehm-Bücherei, A. Ziemsen-Verlag, Wittenberg, 1954, p. 71.

22. *Ibid*., p. 77.

23. SIEGFRIED H. JAECKEL, *Kopffüßer – Tintenfische*. Neue Brehm-Bücherei, A. Ziemsen-Verlag, Wittenberg, 1957, p. 13.

24. LORUS J. MILNE and FRITZ BOLLE, *op. cit.*, p. 281.

25. W. RUTHERFORD, "Phosphorescent Wheel," *The Marine Observer*, 30 (1960), No. 189, p. 128.

26. MARTIN RODEWALD, "Leuchträder des Meeres," *Umschau*, 61 (1961), pp. 177–9.

27. KURT KALLE, "Die rätselhafte und unheimliche Naturerscheinung des explodierenden und des rotierenden Meeresleuchtens," *Deutsche Hydrographische Zeitschrift*, 13 (1960), p. 49.

28. GRUNER, *op, cit.*, p. 41.

29. *Ibid*., p. 26.

30. FRIEDRICH SCHALLER, "Das Licht der Tiere," *Umschau*, 63 (1963), pp. 663–5.

31. TASCHENBERG, *Brehms Tierleben*.

32. VITUS B. DRÖSCHER, *The Mysterious Senses of Animals*. E. P. Dutton & Co., Inc., New York, 1965; Hodder & Stoughton Ltd., London, 1965, p. 46.

33. FRIEDRICH SCHALLER, "Weshalb leuchten die Glühwürmchen?" *Umschau*, 61 (1961), pp. 4–6.

34. FRIEDRICH SCHALLER and H. SCHWALB, "Attrappenversuche mit Larven und Imagines einheimischer Leuchtkäfer," *Verhandlungen Deutsche Zoologische Gesellschaft*, Bonn, 1960, pp. 154–66.

35. WILLIAM D. MCELROY and HOWARD H. SELIGER, "Biological Luminescence," *Scientific American*, Vol. 207, No. 6 (Dec. 1962) pp. 76–89.

36. WILLIAM D. MCELROY and BENTLEY GLASS, "A Symposium on

Light and Life," *Johns Hopkins University Press*, Baltimore, 1961.

37. WHITE, McCAPRA, FIELD, and McELROY, "The Structure and Synthesis of Firefly Luciferin," *Journal of the American Chemical Society*, Vol. 83 (1961), p. 2402.

38. KARL VON FRISCH, *The Dancing Bees: An Account of the Life and Senses of the Honey Bee*. Translated by Dora Ilse. Harcourt, Bruce and World, New York, 1955; Methuen & Co. Ltd., London, 1954.

39. For a more detailed discussion see DIETRICH BURKHARDT, "Untersuchungen an einzelnen Sehzellen," *Umschau*, 64 (1964), pp. 312–13.

40. An insect's vision is sharper than one would imagine from the physiological structure of its eye. This is shown by RUDOLF JANDER and CHRISTIANE VOSS in: "Die Bedeutung von Streifenmustern für das Formensehen der Roten Waldameise," *Zeitschrift Für Tierpsychologie*, 20 (1963), pp. 1–9.

41. HERBERT HERAN, "Wie überwacht die Biene ihren Flug?" *Umschau*, 64 (1964), pp. 299–303.

42. BERNHARD HASSENSTEIN and WERNER RIECHARDT, "Wie sehen Insekten Bewegungen?" *Umschau*, 59 (1959), pp. 302–5.

43. VON FRISCH, *op. cit.*

44. RUDOLF JANDER, "Die Detektortheorie optischer Auslösemechanismen von Insekten," *Zeitschrift für Tierpsychologie*, 21 (1964), pp. 302–7.

45. MARTIN LINDAUER, "Ocellen registrieren den Dämmerungsgrad," *Biologisches Zentralblatt*, 82 (1963), p. 721.

46. EBERHARD DODT, *Mitteilungen der Max-Planck-Gesellschaft*, 1964.

47. J. DE LA MOTTE, "Über die augenunabhängige Lichtwarhrnehmung bei Fischen," review in *Naturwissenschaftliche Rundschau*, 16 (1963), p. 487. Original in *Die Naturwissenschaften*, 50 (1963), p. 363.

48. W. PETRI, "Sehende Finger," *Naturwissenschaftliche Rundschau*, 16 (1963), pp. 407–8.

49. Report, "Eyeless Vision Unmasked," *Scientific American*, Vol. 212, No. 3 (March 1965), p. 57.

50. L. GOETTERT, "Orientierungsmöglichkeiten beim augenlosen Höhlensfisch," *Naturwissenschaftliche Rundschau*, 15 (1962), pp. 56–8.

51. RUDOLF BRAUN, "Zum Lichtsinn augenloser Muscheln," *Zoologisches Jahrbuch, Abteilung Physiologie*, 65 (1954), p. 194.

52. RUDOLF BRAUN, "Der Lichtsinn augenloser Tiere," *Umschau*, 58 (1958), pp. 306–9.

See also WOLFGANG VON BUDDENBROCK, *Vergleichende Physiologie*, Vol. I: *Sinnesphysiologie*. Basel, 1952.

53. VINCENT G. DETHIER and ELIOT STELLAR, *Das Verhalten der Tiere*. Kosmos-Studienbücher, Stuttgart, 1964, p. 15.

54. GERTI DÜCKER, "Farbensehen bei Säugetieren," *Umschau*, 61 (1961), pp. 231–2.

55. Report, "Discriminating Cats," *Scientific American*, Vol. 210, No. 6 (June 1964), p. 59.

56. E. NICKEL, "Vom Farbensinn der Alligatoren," *Zeitschrift für Vergleichende Physiologie*, 43 (1960), p. 37.

57. W. R. A. MUNTZ, "What the Frog's Eye Tells the Frog's Brain," *Journal of Neurophysiology*, November 1962.

58. MARTIN LINDAUER, "Fortschritte der Zoologie, 1963, Vol. 16, Part 1, *Allgemeine Sinnesphysiologie*, pp. 59–140.

59. HANSJOCHEM AUTRUM, "Wie nimmt das Auge Farben wahr?" *Umschau*, 63 (1963), pp. 332–6.

60. EDWARD F. MACNICOL, JR., "Three-Pigment Colour Vision," *Scientific American*, Vol. 211, No. 6 (Dec. 1964), pp. 48–56.

61. GUNNAR SVAETICHIN and EDWARD F. MACNICHOL, "Retinal Mechanisms for Chromatic and Achromatic Vision," *Annual of the New York Academy of Science*, 74 (1958), pp. 385–404.

62. VON FRISCH, *op. cit*.

63. LORUS and MARGERY MILNE, *op. cit*.

64. WULF ENNO ANKEL, "Begegnung mit Limulus," *Natur und Volk*, 88 (1958), pp. 101–10.

65. VITUS B. DRÖSCHER, *op. cit*., p. 240.

Chapter II

1. *Ibid*., pp. 29–32, and T. H. BULLOCK and R. B. COWLES, "Physiology of an Infrared Receptor. The Facial Pit of Vipers," *Science*, Vol. 115, (1952), pp. 541–3.

2. LORUS and MARGERY MILNE, *op. cit*., pp. 90–4.

3. PHILLIP S. CALLAHAN, "Insects Tuned in to Infrared Rays," *New Scientist*, Vol. 23, No. 400 (16 July 1964), pp. 137–8.

4. KONRAD HERTER, *Der Temperatursinn der Tiere*. Neue Brehm-Bücherei, A. Ziemsen-Verlag, Wittenberg, 1962, p. 37.

5. MANFRED ZAHN, "Thermotaktische Orientierung der Schollen, *Umschau*, 63 (1963), p. 711.

6. HERBERT HERAN, "Untersuchungen über den Temperatursinn der Honigbiene," *Zeitschrift für Vergleichende Physiologie*, 34 (1952), pp. 179–206.

7. S. DIJKGRAAF, "Untersuchungen über den Temperatursinn der Fische," *Zeitschrift für Vergleichende Physiologie* 27 (1940),

pp. 587–605; *Zeitschrift für Vergleichende Physiologie*, 30 (1943), p. 252.

8. H. J. FRITH, "Incubator Birds," *Scientific American*, Vol. 201, No. 2 (Aug. 1959), pp. 52–8.

9. PRECHT, CHRISTOPHERSEN, and HENSEL, *Temperatur und Leben*, Springer Verlag, Berlin, 1955.

10. Y. ZOTTERMAN, "Special Senses: Thermal Receptors," *Annual Review of Physiology*, 15 (1953), pp. 357–72.

11. RALPH BUCHSBAUM and LORUS MILNE, *Knaurs Tierreich in Farben – Niedere Tiere*. Droemer-Knaur, Munich 1960, p. 183.

12. HERTER, *op. cit.*, pp. 43–70.

13. T. H. BENZINGER, "The Human Thermostat," *Scientific American*, Vol. 204, No. 1 (Jan. 1961), pp. 134–47.

14. Report, "Cells in the Brain Sensitive to Temperature," *New Scientist*, Vol. 25, No. 428 (28 Jan. 1965), p. 227.

15. RUDOLF THAUER, "Kältesensible Sinneszellen auch im Körpergewebe," *Die Naturwissenschaften*, 1964, No. 4.

16. LORUS and MARGERY MILNE, *op. cit.*, p. 103.

17. RÉMY CHAUVIN, *Tiere unter Tieren*. Scherz-Verlag, Bern, 1964, pp. 120–1.

18. KARL VON FRISCH, *op. cit.*, p. 24.

19. MARTIN LÜSCHER, "Air-Conditioned Termite Nests," *Scientific American*, Vol. 205, No. 1 (July 1961), pp. 138–45.

20. VITUS B. DRÖSCHER, *op. cit.*, pp. 74–8.

21. Another example of an intermediate stage between cold- and warm-blooded animals is provided by the bats. Further detail on this subject is given by Erwin Kulzer in his articles, "Der Thermostat der Fledermäuse," *Natur und Museum*, Vol. 95, No. 8 (Aug. 1965), pp. 331–45, and "Sind die Grossfledermäuse wechselwarme Tiere oder Warmblüter?" in *Umschau*, 63 (1963), pp. 689–92.

22. HERBERT PRECHT, "Anpassungen wechselwarmer Tiere," *Naturwissenschaftliche Rundschau*, 17 (1964), pp. 438–42.

Chapter III

1. HENRY K. BEECHER, *Measurement of Subjective Responses* Oxford University Press, 1959.

2. WILLIAM R. THOMPSON and RONALD MELZACK, "Early Environment," *Scientific American*, Vol. 194, No. 1 (Jan. 1956).

3. PATRICK D. WALL, "Cord Cells Responding to Touch, Damage and Temperature of Skin," *Journal of Neorophysiology*, Vol. 23, No. 2 (March 1960), pp. 197–210.

4. RONALD MELZACK, "The Perception of Pain," *Scientific American*, Vol. 204, No. 2 (Feb. 1961), pp. 41–49.

5. Report, "Ultrasound Exorcises a Phantom Limb," *New Scientist*, Vol. 23, No. 409 (17 Sept. 1964). p. 682.

6. WILLIAM F. HALL, "Sensing Partial Failures – A Step Toward Self-Healing," *Bionics Symposium, 1963*, Contributed Paper Pre-Prints, Wright Patterson Air Force Base, Ohio, p. 259.

Chapter IV

1. WALTER NEUHAUS, "Wieviel Riechsinneszellen besitzen Hunde?", *Umschau*, 55 (1955), p. 421.

2. WALTER NEUHAUS, "Die Fährtenreinheit des Hundes," *Umschau*, 58 (1958), pp. 161–3.

3. WALTER NEUHAUS, "Ist die Riechfähigkeit des Hundes veränderlich?" *Umschau*, 61 (1961), pp. 36–7.

4. KARL P. SCHMIDT and ROBERT F. INGER, *Living Reptiles of the World*. Doubleday, New York, 1957; Hamish Hamilton Ltd., London, 1957, pp. 14–15.

5. Report, "Do Fish Taste Through Their Skin?" *New Scientist*, Vol. 28, No. 470 (18 Nov. 1965), p. 511.

6. IRENÄUS EIBL-EIBESFELDT, *Land of a Thousand Atolls: A Study of Marine Life in the Maldive and Nicobar Islands*. Translated by Gwynne Vevers. International Publications, New York, 1966; MacGibbon & Kee, London, 1965.

7. S. L. SMITH, "Clam-digging Behaviour in the Starfish," *Behaviour*, Vol. 18 (1961), pp. 148–51.

8. NIKO LAAS TINBERGEN, "Von den Vorratskammern des Rotfuches," *Zeitschrift für Tierpsychologie*, 22 (1965), pp. 119–49.

9. J. KLINGLER, "Anziehungsversuche mit CO_2," *Nematologia* (Leiden), 6 (1961), pp. 69–84.

10. HOWARD I. MAIBACH "Insect Attractants," University of California Information Release 1965.

11. R. H. WRIGHT, "Tunes to Which Mosquitoes Dance," *New Scientist*, Vol. 37, No. 590 (28 March 1968), pp. 694–7.

12. HAROLD HEATWOLE, DONALD M. DAVIS, and ADRIAN M. WENNER, "The Behaviour of Megarhyssa," *Zeitschrift für Tierpsychologie*, 19 (1962), pp. 652–64.

13. Report, "Crunching Sounds Rouse These Males," *New Scientist*, Vol. 23, No. 402 (30 July 1964), p. 282.

14. R. C. FISCHER, "A Study in Insect Multiparasitism," *Journal of Experimental Biology*, 38 (1961), pp. 267–75.

15. R. L. DOUTT, "The Biology of Parasitic Hymenoptera," *Annual Review of Entomology*, 4 (1959), pp. 161–82.

16. A. H. KASCHEF, "Sur le comportement de Lariophagus distinguendus," *Behaviour*, 14 (1959), pp. 108–22.

17. FRIEDRICH DÖRBECK, "Die Lachswanderung im nördlichen Fernosten," *Natur und Volk*, 85 (1955), pp. 391–9.

18. J. R. BRETT, "The Swimming Energetics of Salmon," *Scientific American*, Vol. 213, No. 2 (Aug. 1965), pp. 80–5.

19. W. A. CLEMENS, R. E. FOERSTER, and A. L. PITCHARD, "Migration and Conservation of Salmon," *Publications of the American Association for the Advancement of Science*, 8 (1939), pp. 51–59.

20. L. R. DONALDSON and G. H. ALLEN, *Transactions of the American Fisheries Society*, 87 (1957), p. 13.

21. ARTHUR D. HASLER, "Wegweiser für Zugfische, "*Naturwissenschaftliche Rundschau*, 15 (1962), pp. 302–10.

22. HARALD TEICHMANN, "Das Riechvermögen des Aales," *Naturwissenschaften*, 44 (1957), p. 242.

23. ADOLF BUTENANDT: "Über Wirkstoffe des Insektenreiches," *Naturwissenschaftliche Rundschau*, 8 (1955), pp. 457–64.

24. ERICH HECKLER, "Sexuallockstoffe – hochwirksame Parfüms der Schmetterlinge," *Umschau* 59 (1959), pp. 465–7, 499–502.

25. VITUS B. DRÖSCHER, *The Mysterious Senses of Animals.* E. P. Dutton & Co., Inc., New York, 1965; Hodder & Stoughton, London, 1965, pp. 136–7.

26. GÜNTHER STEIN, "Der Sexuallockstoff von Hummelmännchen," *Umschau*, 64 (1964), p. 54.

27. MARTIN LINDAUER, *Fortschritte der Zoologie*, Vol. 16 (1963), Part 1, *Orientierung im Raum*, p. 100.

28. B. KULLENBERG, "Field Experiments with Chemical Sexual Attractants," *J. Zool. Bidr. fran. (Uppsala)*, 31, (1956), pp. 253–4.

29. A. BUTENANDT, R. BECKMANN, D. STAMM, and E. HECKER, "Der Sexuallockstoff des Seidenspinners," *Zeitschrift für Naturforschungen* 14b (1959), p. 283.

30. MARTIN JACOBSON and MORTON BEROZA, "Insect Attractants," *Scientific American*, Vol. 211, No. 2 (Aug. 1964), pp. 20–7.

31. Report, "Versagen eines Sexuallockstoffes durch geringe Verunreinigungen," *Umschau*, 65 (1965), p. 720.

32. KARL V. FRISCH, "Über einén Schreckstoff der Fischhaut und seine biologische Bedeutung," *Zeitschrift für Vergleichende Physiologie*, 29 (1941), pp. 46–145.

33. WOLFGANG PFEIFFER, "Die Schreckreaktion der Fische," *Umschau*, 65 (1965), pp. 401–5.

34. ERWIN KULZER, "Neuere Untersuchungen über Schreck- und Warnstoffe im Tierreich." *Naturwissenschaftliche Rundschau*, 12 (1959), pp. 296–302.

35. F. SCHUTZ, "Vergleichende Untersuchungen über die Schreckreaktion bei Fischen," *Zeitschrift für Vergleichende Physiologie*, 38 (1956), pp. 84–135.

36. ROLF HENNIG: "Über einige Verhaltensweisen des Rehwildes in freier Wildbahn," *Zeitschrift für Tierpsychologie*, 19 (1962), pp. 223–9.

37. ROLF HENNIG, "Über das Revierverhalten der Rehböcke," *Zeitschrift für Jagdwissenschaft*, 8 (1962), pp. 61–81.

38. IRENÄUS EIBL-EIBESFELDT, "Angeborenes und Erworbenes im Verhalten einiger Säuger," *Zeitschrift für Tierpsychologie*, 20 (1963), p. 733.

39. Report, "Why Rabbits Rub Their Chins," *New Scientist*, Vol. 26, No. 438 (8 April 1965), p. 78.

40. BERNHARD GRZIMEK, *Wir lebten mit den Baule*. Verlag Ullstein, Berlin, 1963, p. 38.

41. GÜNTER TEMBROCK, *Grundlagen der Tierpsychologie*. Akademie-Verlag, Berlin, 1963, p. 164.

42. LINDOVER, *op. cit.*, p. 110.

43. KONRAD LORENZ, *On Aggression*, Methuen, London, 1966.

44. PETER KARLSON and ADOLF BUTENANDT, "Pheromones (Ectohormones) in Insects," *Annual Review of Entomology*, Vol. 4 (1959), pp. 39–58.

45. P. KARLSON and M. LÜSCHER, "Pheromone," *Die Naturwissenschaften*, 1959, pp. 63–4.

46. EDWARD O. WILSON, "Pheromones," *Scientific American*, Vol. 208, No. 5 (May 1963), pp. 100–14.

47. G. H. SCHMIDT, "Pheromone als Sputstoffe bei Ameisen," *Naturwissenschaftliche Rundschau*, 18 (1965), pp. 197–8.

48. D. BOTSCH, "Mathematische Analyse der Pheromonwirkung bei Insekten," *Naturwissenschaftliche Rundschau* 17 (1964), p. 149.

49. Report: "Termiten geben Klopfzeichen," *Umschau*, 64 (1964), p. 155

50. PETER KAISER, "Hormonalorgane steuern die Kastentwicklung der Termiten," *Umschau*, 56 (1956), pp. 651–3.

51. H. SCHERF, "Sozialwirkstoffe bei Termiten," *Naturwissenschaftliche Rundschau*, 15 (1962), p. 322.

52. J. PAIN, "Sur la phéromone des reines d'abeilles et ses effets physiologiques." *Ann. Abeille*, 4 (1961), pp. 73–152.

53. Report, "Mosquitoes Succumb to Queen Substance," *New Scientist*, Vol. 27, No. 453 (July 1965), p. 219.

54. C. B. WILLIAMS, *Insect Migration*. William Collins Sons & Co. Ltd., London, 1958, p. 78.

55. ADOLF REMANE, "Das soziale Leben der Tiere," *Rowohlts deutsche Enzyklopädie*, Nr. 97, pp. 9–11.

56. RÉMY CHAUVIN, *op, cit.*, pp. 150–7.

57. T. T. MACAN, "Self-controls on Population Size," *New Scientist*, Vol. 28, No. 474 (Dec. 1965), pp. 801–3.

58. V. C. WYNNE-EDWARDS, *Animal Dispersion in Relation to Social Behaviour*. Oliver and Boyd, Edinburgh, 1962.

59. HELEN M. BRUCE and A. S. PARKES, "Olfactory Stimuli in Mammalian Reproduction," *Science*, Vol. 134, No. 3485 (Oct. 1961), pp. 1049–1054.

60. Report, "Identifying People by their Smell," *New Scientist*, Vol. 28, No. 472 (Dec. 1965), p. 650.

61. DIETRICH SCHNEIDER, "Neue Experimente zum Geruchsproblem," *Sandorama*, June 1965, pp. 22–3.

62. JOHN E. AMOORE, JAMES W. JOHNSTON, and MARTIN RUBIN, "The Stereochemical Theory of Odor," *Scientific American*, Vol. 210, No. 2 (Feb. 1964), pp. 42–9.

63. DIETRICH SCHNEIDER, "Vergleichende Rezeptorphysiologie am Beispiel der Riechorgane von Insekten," *Jahrbuch 1963 der Max-Planck-Gesellschaft*, pp. 149–177.

64. DIETRICH SCHNEIDER, VEIT LACHER, and KARL-ERNST KAISSLING, "Die Reaktionsweise und das Reaktionsspektrum von Riechzellen bei Antheraea pernyi," *Zeitschrift für Vergleichende Physiologie*, 48 (1964), pp. 632–62.

65. VEIT LACHER, "Elektrophysiologische Untersuchungen an enizelnen Rezeptoren für Geruch, Kohlendioxyd, Luftfeuchtigkeit und Temperatur auf den Antennen der Arbeitsbiene und der Drohne," *Zeitschrift für Vergleichende Physiologie*, 48 (1964), pp. 587–623.

66. C. V. EULER and U. SÖDERBERG, "Medullary Chemosensitive Receptors," *Journal of Physiology* London 118, (1952), pp. 545–54.

Chapter V

1. IVAN T. SANDERSON, *Knaurs Tierreich in Farben – Säugetiere*. Droemersche Verlagsanstalt, München 1956, p. 54.

2. FRANZ PETER MÖHRES and E. KULZER, *Zeitschr für Vergleichende Physiologie*, 38 (1956), p. 1.

3. DONALD R. GRIFFIN, "Echo-Ortung der Fledermäuse," *Naturwissenschaftliche Rundschau*, 15 (1962), pp. 169–73.

4. DONALD R. GRIFFIN, *Listening in the Dark*. Yale University Press, New Haven, 1958, p. 413.

5. ANTON KOLB, "Wie orientieren sich Fledermäuse während des Fressens?" *Umschau*, 65 (1965), pp. 334–5.

6. FRANZ PETER MÖHRES, "Bildhören – eine neuentdeckte Sinnesleistung der Tiere," *Umschau*, 60 (1960), pp. 673–8.

7. ANTON KOLB, "Jagen Fledermäus nur im Fluge?" *Umschau*, 59 (1959), pp. 334–5.

8. HEINRICH HERTEL, *Struktur – Form – Bewegung*, in the series *Biologie und Technik*, Kraußkopf Verlag, Mainz, 1963, pp. 23–4.

9. KENNETH D. ROEDER, "Moths and Ultrasound," *Scientific American*, Vol. 212, No. 4 (April 1965), pp. 94–102.

10. KENNETH D. ROEDER and ASHER E. TREAT, "The Detection and Evasion of Bats by Moths," *American Scientist*, Vol. 49, No. 2 (June 1961), pp. 135–48.

11. HANS KIETZ, personal communication to the author.

12. DOROTHY C. DUNNING and KENNETH D. ROEDER, "Moth Sounds and the Insect-Catching Behaviour of Bats," *Science*, Vol. 147, No. 3654 (Jan. 1965), pp. 173–4.

13. DOROTHY C. DUNNING, "Moths Are Warning Bats," *Zeitschrift für Tierpsychologie*, Vol. 25, No. 2, pp. 129–38.

14. HUBERT FRINGS and MABLE FRINGS, "Sound Against Insects," *New Scientist*, Vol. 26, No. 446 (June 1965), pp. 634–7.

15. HANS HASS, personal communication to the author.

16. For further details see S. RAUCH: "Die Ionenverteilung im Innenohr," *Umschau*, 65 (1965), pp. 171–2.

17. HANS SCHNEIDER, "Auch Fische haben eine Sprache," *Umschau*, 64 (1964), pp. 166–70.

18. IRENÄUS EIBL-EIBESFELDT, *Land of a Thousand Atolls: A Study of Marine Life in the Maldive and Nicobar Islands*. Translated by Gwynne Vevers. International Publishers, New York, 1966; MacGibbon & Kee, London, 1965.

19. V. C. WYNNE-EDWARDS, *op. cit.*, pp. 71, 196, 201–2, 337–8.

20. HANS SCHNEIDER, *op. cit.*

21. SVEN DIJGRAAF, "The Functioning and Significance of the Lateral-line Organs," *Biological Review*, 38 (1963), pp. 51–105.

22. J. SCHWARTZKOPFF, "Die Stufenleiter des Hörens," *Umschau*, 60, (1960), pp. 4–7.

23. CLAUS TIMM, "Ultraschallhören," *Experientia*, 1950, p. 3571.

24. E. ZWICKER, "Funktionsmodelle bei der Erforschung des Gehörs," *Umschau*, 63, (1963), pp. 698–701.

25. MARK R. ROSENZWEIG, "Auditory Localization," *Scientific American*, Vol. 205, No. 4 (Oct. 1961), pp. 132–42.

26. WERNER ENDRES, "Automatische Spracherkennung," *Umschau*, 66 (1966), pp. 152–7.

27. W. C. DERSCH, "A Decision Logic for Speech Recognition," *IBM-Technical Report*, December 1961.

28. H. G. TILLMANN, *Konstruktion eines Automaten zur Identifikation von Wortsignalen*. International report of the Institute for Research in Phonetics and Communication, University of Bonn, 1964.

27. H. KUSCH, "Automatische Erkennung gesprochener Zahlen," *Nachrichtentechnische Zeitschrift*, 18 (1965), Vol. 2, pp. 57–62.

Chapter VI

1. ERNST KILIAN, "Wie verhalten sich Tiere bei Erdbeben?" *Naturwissenschaftliche Rundschau*, 17 (1964), pp. 135–9.

2. WOLFDIETRICH KÜHME, "Verhaltensstudien am maulbrütenden und am nestbauenden Kampffisch," *Zeitschrift für Tierpsychologie*, 18 (1961), pp. 33–55.

3. LORUS and MARGERY MILNE, *The Senses of Animals and Men*. Atheneum, New York, 1962, p. 33.

4. ERWIN TRETZEL, "Die Sprache bei Spinnen," *Umschau*, 63 (1963), pp. 372–6, 403–7.

5. Report, "Buzzing the Queen," *Scientific American*, Vol. 207, No. 6 (Dec. 1962), pp. 70–1.

6. HARALD ESCH, "Über die Schallerzeugung beim Werbetanz der Honigbiene," *Zeitschrift für Vergleichende Physiologie*, 45 (1961), pp. 1–11.

7. HARALD ESCH, "Auch Lautäusserungen gehören zur Sprache der Bienen," *Umschau*, 62 (1962), pp. 293–6.

8. Report, "Device Helps the Blind to See Print," *New Scientist*, Vol. 24, No. 411 (Oct. 1964), p. 10.

9. R. DARCHEN, "La construction sociale chez Apis mellifica," *Insectes sociaux*, 3 (1956).

10. RÉMY CHAUVIN, *op. cit.*, p. 84.

11. D. MERRILL, "Why Caddis-Worms Stop Building," *New Scientist*, Vol. 26, No. 445 (May 1965), p. 589.

12. W. RATHMAYER, "Das Paralysierungsproblem beim Bienenwolf," *Zeitschrift für Vergleichende Physiologie*, 45 (1962), pp. 413–62.

13. LORUS and MARGERY MILNE, *op. cit.*, p. 16.

14. OTTO KOENIG, *Kif-Kif. Menschliches und Tierisches zwischen Sahara und Wilhelminenberg*. Wollzeilenverlag, Wein, 1962, p. 201.

15. D. BURKHARDT and G. SCHNEIDER, *Zeitschrift für Naturforschung*, 12 b (1957), p. 139.

16. VITUS B. DRÖSCHER, *op. cit.*, pp. 40–1.

17. GEORG BIRUKOW, "Die Windorientierung des Mistkäfers," *Zeitschrift für Tierpsychologie*, 15 (1958), p. 265.

18. HERBERT HERAN, "Wie überwacht die Biene ihren Flug?" *Umschau*, 64 (1964), pp. 299–303.

19. VOLKER NEESE, "Zur Funktion der Augenborsten bei der Honigbiene," *Zeitschrift für Vergleichende Physiologie*, 49 (1965), pp. 543–85.

Chapter VII

1. DETLEF BÜCKMANN, "Nehmen Insekten die Schwerkraft wahr?" *Umschau*, 56 (1956), pp. 309–11.

2. U. BÄSSLER, "Versuche zur Orientierung der Stechmücken," *Zeitschrift für Vergleichende Physiologie*, 41 (1958), p. 300.

3. MARTIN LINDAUER and J. O. NEDEL, *Zeitschrift für Vergleichende Physiologie*, 42 (1959), p. 334.

4. HUBERT MARKL, "Wie orientieren sich Ameisen nach der Schwerkraft?" *Umschau*, 65 (1965), pp. 185–8.

5. HELMUTH DECHER, "Neue Erkenntnisse über den menschlichen Gleichgewichtsapparat, *Umschau*, 65 (1965), pp. 738–40.

6. HERMAN A. WITKIN, "The Perception of the Upright," *Scientific American*, Vol. 200, No. 2 (Feb. 1959), pp. 50–6.

7. HERMAN A. WITKIN, *Personality Through Perception. An Experimental and Clinical Study*. Harper and Brothers, New York, 1954.

8. W. B. CANNON and A. L. WASHBURN, *American Journal of Physiology*, Vol. 29 (1912), pp. 441–54.

9. C. H. KAYSER, "Wie entsteht Hunger?" *Umschau*, 65 (1965), pp. 129–32.

10. J. MAYER, *Annual of the New York Academy of Science*, 63 (1955), pp. 15–43.

11. ARTHUR JORES, "Ist Fettsucht eine Krankheit?" *Wiesbadener Symposium der Deutschen Gesellschaft für Innere Medizin*, 1965.

12. M. HERTZ, "Eine Bienendressur auf Wasser," *Zeitschrift für Vergleichende Physiologie*, 21 (1935), pp. 463–7.

13. Report, "A Pressure Gauge in the Kidney?" *New Scientist*, Vol. 28, No. 471 (Nov. 1965), pp. 561–2.

14. BENJAMIN W. ZWEIFACH, "The Microcirculation of the Blood," *Scientific American*, Vol. 200, No. 1 (Jan. 1959), pp. 54–60.

15. CLEMENT A. SMITH, "The First Breath," *Scientific American*, Vol. 209, No. 4 (October 1963), pp. 27–35.

16. HANS WINTERSTEIN, "50 Jahre Reaktionstheorie der Atmung," *Naturwissenschaftliche Rundschau*, 14 (1961), pp. 413–15.

17. KEITH R. PORTER and CLARA FRANZINI-ARMSTRONG, "The Sacroplasmic Reticulum," *Scientific American*. Vol. 212, No. 3 (March 1965), pp. 73–80.

18. G. J. V. NOSSAL, "How Cells Make Antibodies," *Scientific American*, Vol. 211, No. 6 (Dec. 1964), pp. 106–14.

19. H. W. LISSMANN, "On the Function and Evolution of Electric Organs in Fish," *The Journal of Experimental Biology*, 35 (1958), pp. 156–91.

20. H. W. LISSMANN, "Electric Location by Fishes," *Scientific American*, Vol. 208, No. 3 (March 1963), pp. 50–9.

21. A summation of present knowledge in this field will be found in HARRY GRUNDFEST, "Electric Fishes," *Scientific American*, Vol. 203, No. 4 (Oct. 1960), pp. 115–24.

22. FRANZ PETER MÖHRES, "Die elektrischen Fische," *Natur und Volk*, 91 (1961), pp. 1–13.

23. WILHELM HARDER, "Elektrische Fische," *Umschau*, 65 (1965), pp. 467–73, 492–496.

Chapter VIII

1. E. THOMAS GILLIARD, *Living Birds of the World*. Doubleday and Company, Garden City, New York; Hamish Hamilton Ltd., London, 1958, p. 207.

2. Report, "The Remarkable Time-Table of the Mutton-birds," *New Scientist*, Vol. 23, No. 401 (July 1964), p. 203.

3. ERNST SCHÜZ, *Vom Vogelzug*. Verlag. Dr. Paul Schöps, Frankfurt am Main, 1952, p. 39.

4. H. RITTINGHAUS, "Flughöhenvon Zugvögeln," *Die Vogelwarte*, 19 (1957), Part 2, p. 90.

5. ERICH DAUTERT, *Auf Walfgang und Robbenjagd im Südatlantik*.

6. HANNO CILIAX, personal communication to the author.

7. E. J. SLIJPER, W. L. VAN UTRECHT, and C. NAAKTGEBOREN, "Ergebnisse der Walforschung an Hand von Schiffsbeobachtungen," *Umschau*, 65 (1965), pp. 774–9.

8. E. J. SLIJPER, *Riesen des Meeres, eine Biologie der Wale und Delphine*. Verständliche Wissenschaft, Springer-Verlag. Berlin, 1962.

9. Report, "Isotopes for Tracing Whale Movements?" *New Scientist*, Vol. 24, No. 422 (Dec. 1964), p. 770.

10. Report, "Transatlantische Reise eines markierten Kabeljaus," *Umschau*, 63 (1963), p. 320.

11. WOLFGANG PFEIFFER, "Die geruchliche und optische Orientierung der Fische," *Dragoco Report*, 1965, pp. 231–41.

12. Report, "Thune überqueren die Ozeane," *Umschau*, 60 (1960), p. 306.

13. G. HEMPEL, "Schwankungen in den skandinavischen Herings-
    beständen," *Umschau*, 61 (1961), pp. 758–9.
14. ERNA PINNER, "Die lange Reise des Dornhais," *Naturwissen-
    schaftliche Rundschau*, 18 (1965), pp. 205–6.
15. D. W. TUCKER, *Nature*, Vol. 183 (1959), p. 495.
16. Report, "Wandern unsere Aale ins Sargasso-Meer?" *Umschau*,
    60 (1960), p. 594.
17. CARRINGTON B. WILLIAMS, *Insect Migration*. William Collins
    Sons & Co. Ltd., London, 1958, pp. 111–12.
18. C. G. JOHNSON, "The Aerial Migration of Insects," *Scientific
    American*, Vol. 209, No. 6 (Dec. 1963), pp. 132–8.
19. K. BURMANN, "Massenflüge des grauen Lärchenwicklers,"
    Anzeiger für Schädlingskunde, 38 (1965), pp. 4–7.

Chapter IX

1. E. R. BAYLOR and F. E. SMITH, "The Orientation of Cladocera
    to Polarized Light, "*American Naturalist*, 87 (1953), p. 97
2. R. BAINBRIDGE and T. H. WATERMAN, "Polarized Light and the
    Orientation of Two Marine Crustacea," *Journal of Experi-
    mental Biology*, 34 (1957), p. 342.
3. MARIANNE GEISLER, "Untersuchungen zur Tagesperiodik des
    Mistkäfers," *Zeitschrift für Tierpsychologie*, 18 (1961), pp. 389–
    420.
4. For further details see ERWIN BÜNNING, *Die physiologische
    Uhr*, Springer Verlag, Berlin, 2nd edition, 1963.
5. GEORG BIRUKOW, "Lichtkompassorientierung beim Wasser-
    läufer," *Zeitschrift für Tierpsychologie*, 13 (1956), pp. 463–84.
6. L. PARDI, "Modificazione sperimentale della direzione di fuga
    negli anfipodi ad orientatmento solare," *Zeitschrift für Tier-
    psychologie*, 14 (1957), pp. 261–75.
7. F. PAPI and P. TONGIORGI, "Innate and Learned Components
    in the Astronomical Orientation of Wolf Spiders," *Ergebnisse
    der Biologie*, 26 (1963), pp. 259–80.
8. KARL V. FRISCH, *Tanzsprache und Orientierung der Bienen*,
    Springer-Verlag, Berlin, 1965, pp. 134–6, 367.
9. J. REIMANN, "Die Sonnenorientierung der Waldameise,"
    unpublished thesis, Freiburg in Breisgau, 1964, cited by
    RUDOLF JANDER, "Die Hauptentwicklungsstufen der Licht-
    orientierung bei den tierischen Organismen,' *Naturwissen-
    schaftliche Rundschau*, 18 (1965), pp. 318–24.
10. KARL VON FRISCH, *op. cit.*, pp. 384–476.
11. T. H. GOLDSMITH and D. E. PHILPOTT, "The Microstructure of
    the Compound Eyes of Insects," *Journal of Biophysical and
    Biochemical Cytology*, 3 (1957), pp. 429–38.

12. T. H. GOLDSMITH, "Fine Structure of the Retinulae in the Compound Eye of the Honeybee", *Journal of Cell Biology*, 14 (1962), pp. 489–94.

13. A. C. PERDECK, "Two Types of Orientation in Migrating Starlings ' *Ardea* (Leiden) 46 (1958), p. 1–37.

14. KLAUS SCHMIDT-KOENIG, "Über die Orientierung der Vögel," *Die Naturwissenschaften*, 51 (1964), pp. 423–31.

15. HANS FREIHERR GEYR VON SCHWEPPENBURG, "Zur Terminologie und Theorie der Leitlinie," *Journal für Ornithologie*, 104 (1963), pp. 191–204.

16. GUSTAV KRAMER, "Die Sonnenorientierung der Vögel," *Verhandlungen der Deutschen Zoologischen Gesellschaft in Freiburg* 1952, Leipzig 1953, pp. 72–84.

17. HANS LÖHRL, "Vom Orientierungssinn der Tiere," *Sandosz-Panorama*, Nov. 1964.

18. G. V. T. MATTHEWS, *Bird Navigation*. Cambridge University Press, Cambridge 1955.

19. C. J. PENNYCUICK, *Journal of Experimental Biology*, 37 (1960), p. 573.

20. KLAUS SCHMIDT-KOENIG, "Neue Versuche zum Orientierungsvermögen von Brieftauben, "*Umschau*, 65 (1965), 502–7.

21. WOLF ENGELS, MATHILDE ESSER, and HINRICH RAHMANN, "Anlockung nächtlich ziehender Kraniche durch Grobstadtlichter," *Die Vogelwarte*, 22 (1964), pp. 177–8.

22. FRANZ and ELEONORE SAUER, "Zugvögel als Navigatoren," *Naturwissenschaftliche Rundschau*, 13 (1960), pp. 88–95. See also VITUS B. DRÖSCHER, *op. cit*., pp. 172–80.

23. STEPHEN T. EMLEN, "Migratory Orientation in the Indigo Bunting, *Passerina cyanea*," *Auk*, Vol. 84, pp. 309–42, 463–89.

24. KARL CLEVE, "Der Anflug der Schmetterlinge an künstliche Lichtquellen," *Mitteilungen der Deutschen Entomologischen Gesellschaft*, 23 (1964), pp. 66–76.

25. FRIEDRICH WILHELM MERKEL and WOLFGANG WILTSCHKO, "Magnetismus und Richtungsfinden zugunruhiger Rotkehlchen," *Die Vogelwarte*, 23 (1965), pp. 71–7.

26. FRIEDRICH WILHELM MERKEL, HANS GEORG FROMME and WOLFGANG WILTSCHKO, "Nichtvisuelles Orientierungsvermögen bei nächtlich zugunruhigen Rotkehlchen," *Die Vogelwarte*, 22 (1964), pp. 168–73.

27. HANS GEORG FROMME, "Untersuchungen über das Orientierungsvermögen nächtlich ziehender Kleinvögel," *Zeitschrift für Tierpsychologie*, 18 (1961), pp. 205–20.

28. A. L. DEVRIES and D. E. WOHLSCHLAG, *Science*, 145 (1964), p. 292.

29. JACQUES BOVET, "Ein Versuch, wilde Mäuse unter Ausschluss optischer, akustischer und osmischer Merkmale auf Himmelsrichtungen zu dressieren," *Zeitschrift für Tierpsychologie*, 22 (1965), pp. 839–59.

30. J. O. HÜSING, F. STRUSS, and W. WEIDE, *Die Naturwissenschaften*, 47 (1960), pp. 22–3.

31. GÜNTHER BECKER, "Wirkung magnetischer Felder auf Insekten," *Zeitschrift für angewandte Entomologie*, 54 (1964), pp. 75–88.

32. GÜNTHER BECKER, "Eine Magnetfeldorientierung bei Termiten,' *Die Naturwissenschaften*, 50, (1963), p. 455.

33. F. SCHNEIDER, "Beeinflussung der Aktivät des Maikäfers durch Veränderung der gegenseitigen Lage magnetischer und elektrischer Felder," *Mitteilungen der Schweizerischen Entomologischen Gesellschaft*, 33 (1961), pp. 232–7.

34. OTTO KOEHLER, review of H. J. AUTRUM, "Ergebnisse der Biologie – Orientierung der Tierre," 26 (1963), Springer-Verlag, Berlin, pp. 135–46, in *Zeitschrift für Tierpsychologie*, 20 (1963), p. 762.

35. F. A. BROWN, M. F. BENNETT, and H. M. WEBB, "A Magnetic Compass Response of an Organism," *Biological Bulletin*, 119 (1960), pp. 65–74.

36. LESTER TALKINGTON, "Magnetic-Force Theory on Migration Supported," *Medical Tribune*, Feb. 1965.

37. HERMANN REICH, personal communication to the author.

Index

Panther Science

The Language of Life
An Introduction to the Science of Genetics
George and Muriel Beadle   42½p

Genetics, a relatively new science, is concerned with heredity and variations from it, and its significance is that for good or for ill it may soon be in a position to modify the biology of the human being. Because it is a new discipline its way of thought and its language have tended to baffle most readers. This present book is one of the first to deal with the vital problem of communication.

'Dr. Beadle, a geneticist whose work earned him the Nobel Prize in 1958, explained each part of the subject to his wife who had no scientific training, and it was she who actually wrote the book. This has removed the major barrier of language that exists between scientist and laymen. The terms used are clearly explained, helped by a free use of metaphor and a simple style'
Times Educational Supplement

Panther Science

The Environment Game
Nigel Calder $42\frac{1}{2}$p

The author, until recently editor of *New Scientist*,
argues that the agricultural method of producing
food has become too wasteful of the world's
land areas. We must plan to produce our food
as we produce motor cars or clothing – in
factories. An abundance of food
photo-synthetically produced – and, hand in
hand, an abundance of reverted land to play
with. Present agricultural areas will return to their
pristine condition; vast tracts of a splendidly
re-invigorated Mother Earth will become our
playground. Not, as today, faraway, exclusive,
expensive playgrounds for a small minority – but
a world for all of us.

'Any solution to the problems posed by the
present expansion of the world's population and
the still accelerating productive capacity of
technology in other directions must be to some
extent Utopian. Mr. Calder's Utopia is curious,
original and logical'
Times Educational Supplement

SOME PANTHER AUTHORS

Norman Mailer
Jean-Paul Sartre
Len Deighton
Henry Miller
Georgette Heyer
Mordecai Richler
Gerard de Nerval
James Hadley Chase
Juvenal
Violette Leduc
Agnar Mykle
Isaac Asimov
Doris Lessing
Ivan Turgenev
Maureen Duffy
Nicholas Monsarrat
Fernando Henriques
B. S. Johnson
Edmund Wilson
Olivia Manning
Julian Mitchell
Christopher Hill

Robert Musil
Ivy Compton-Burnett
Chester Himes
Chaucer
Alan Williams
Oscar Lewis
Jean Genet
H. P. Lovecraft
Anthony Trollope
Robert van Gulik
Louis Auchincloss
Vladimir Nabokov
Colin Spencer
Alex Comfort
John Barth
Rachel Carson
Simon Raven
Roger Peyrefitte
J. G. Ballard
Mary McCarthy
Kurt Vonnegut
Alexis Lykiard